The Custodians of Biodiversity
Sharing Access and Benefits to Genetic Resources

Manuel Ruiz and Ronnie Vernooy

International Development Research Centre
Ottawa • Cairo • Dakar • Montevideo • Nairobi • New Delhi • Singapore

First published 2012
by Earthscan
2 Park Square, Milton Park, Abingdon, Oxon OX14 4RN

Simultaneously published in the USA and Canada
by Earthscan
711 Third Avenue, New York, NY 10017

Earthscan is an imprint of the Taylor & Francis Group, an informa business

© 2012 International Development Research Centre

Copublished with the
International Development Research Centre
PO Box 8500, Ottawa, ON K1G 3H9, Canada
info@idrc.ca/www.idrc.ca

British Library Cataloguing in Publication Data
A catalogue record for this book is available from the British Library

Library of Congress Cataloging in Publication Data
The Custodians of biodiversity : sharing access and benefits to genetic
resources / [edited by] Manuel Ruiz and Ronnie Vernooy.
    p. cm.
Includes bibliographical references and index.
1. Crops—Germplasm resources—Economic aspects. 2. Crops—
Germplasm resources—Government policy. I. Ruiz Muller, Manuel.
II. Vernooy, Ronnie, 1963-
SB123.3. C87 2012
338.1'62—dc23   2011021722

ISBN: 978-1-84971-451-8 (hbk)
ISBN: 978-0-203-15603-2 (ebk)

Typeset in Times New Roman
by RefineCatch Limited, Bungay, Suffolk

Contents

List of contributors

Kamalesh Adhikari is research director and editor of *Trade Insight* magazine at South Asia Watch on Trade, Economics and Environment (SAWTEE), a regional network of 11 South Asian NGOs. He has been working on trade, biodiversity, intellectual property rights and farmers' rights issues and has a number of publications in these areas. Mr Adhikari is also a member of the faculty at the National College of Development Studies, Kathmandu. Currently, he sits on the Curriculum Development Committee of the Institute of Agriculture and Animal Science, Nepal. He has been invited to provide advice to several policy committees of the Government of Nepal and has worked as a consultant for several national, regional and international organizations including the Government of Nepal, United Nations Development Programme (UNDP) Nepal, UNDP Regional Centre in Bangkok, and the Food and Agriculture Organization of the United Nations.

Adnan Al-Yassin obtained his MSc and PhD degrees in barley breeding at the University of Jordan in Amman. From 1992 to 2010, he worked at the National Center for Agricultural Research and Extension as a barley researcher. During this period, he led the cereal research program and then directed field crops research (2003–10). In addition to serving as an external examiner for some MSc dissertations, he has coordinated several externally funded projects. Recently, he joined the barley team at the International Center for Agricultural Research in Dry Areas (ICARDA).

Salvatore Ceccarelli was a forage and barley breeder at ICARDA till 2006 and is now a consultant for the Biodiversity and Gene Management Program. He has been involved in participatory plant breeding since 1995.

Teresa Cruz Sardiñas works for the Environmental Direction at the Ministry of Science, Technology and the Environment, Havana, Cuba. Mrs Cruz has been a representative for Cuba to the Convention on Biological Diversity and has participated in a series of official meetings and workshops regarding access to genetic resources, biosafety and biodiversity in general.

Alessandra Galié is a social and gender analysis specialist at ICARDA, Syria, and a PhD candidate at Wageningen University, the Netherlands. Her current research involves social impact assessment of evolutionary participatory plant breeding and women farmers with a focus on empowerment, knowledge and seed governance.

Omar Gallardo is an agronomist and coordinator of the Fundación para la Investigación Participativa con Agricultores de Honduras (FIPAH) in the region of Jesús de Otoro, Intibucá. He obtained his degree in 1999 from the Regional Center for the Littoral Atlantic (CURLA) at the National Autonomous University of Honduras, La Ceiba.

Marvin Gomez is an agronomist and coordinator of FIPAH in Yorito, Sulaco and Victoria. He obtained his degree in 2004 from the Regional Center for the Littoral Atlantic (CURLA) at the National Autonomous University of Honduras, La Ceiba.

Stefania Grando is a barley breeder at ICARDA, Syria. She has more than 25 years of experience in barley breeding. She is actively involved in research to enhance and stabilize productivity and the quality of barley in developing countries with a focus on breeding varieties for poor farming communities.

Sally Humphries is an associate professor in the Department of Sociology and Anthropology at the University of Guelph, Ontario, Canada. She is also the university's director of international development studies. She obtained her PhD in sociology in 1989 at York University, Toronto. She is a founding member of FIPAH and has been supporting research with hillside farmers in Honduras since 1993.

José Jiménez is an agronomist and president of FIPAH. In 1981, he obtained his degree from CURLA at the National Autonomous University of Honduras. Between 1981 and 1993, before becoming involved in participatory research with hillside farmers, he worked for the Honduran Ministry of Agriculture. During that time, among other positions, he was director of the Honduran bean research program.

Li Jingsong earned a BSc degree in sociology from China Agriculture University and a master's degree in environmental management from Wageningen University in the Netherlands. She works as a senior research assistant at the Center for Chinese Agricultural Policy, Chinese Academy of Sciences, Beijing, and has a special interest in rural development and environmental management. In September 2007, she began PhD studies at Wageningen University in the Netherlands.

Yasmin Mustafa is an agricultural economist at ICARDA. Her current research is on impact assessment of participatory and conventional plant breeding, economic evaluations and risk analysis of improved technologies and management practices. She is also estimating the expected adoption profile and potential impact of these programs and technologies on systems productivity, farm income and farmers' well-being.

Bikash Paudel, program officer in socioeconomics at Local Initiatives for Biodiversity, Research and Development (LI-BIRD), is the coordinator of the

Policy and Social Research for Wider and Inclusive Impacts (PSR) Program. He has an MScAg (Agriculture Economics) from G.B. Pant University for Agriculture and Technology, Uttrakhand, India. He has been actively involved in action research and advocacy in agrobiodiversity policy, community-based biodiversity management, farmers' rights and ABS for about four years and has been invited to provide advice to many committees and meetings related to seed and agrobiodiversity in Nepal.

Humberto Ríos Labrada graduated in 1984 as an agronomist from the Higher Pedagogical Institute for Technical and Professional Education, Havana, Cuba. In 1999, he obtained his PhD in agricultural sciences. He has been involved in participatory research in Ghana, Holland, Venezuela, Mexico, Nicaragua, Honduras, Colombia and Cuba, for which he received the Rural Innovation Award of the International Agriculture Centre, the Netherlands. In 2010, he received the Goldman Award from the Goldman Foundation, USA, for his pioneering work on introducing and expanding participatory plant breeding in Cuba. Currently, he is coordinating a national program on local innovation in agriculture in Cuba. He has been a lecturer in plant breeding and experimental designs for the last 12 years and also advises MSc and PhD students.

Manuel Ruiz is a lawyer and a graduate of the Catholic University of Peru. He is currently the director of the Program of International Affairs and Biodiversity of the Peruvian Society for Environmental Law. Manuel specialized in biodiversity, intellectual property and indigenous peoples related issues and has undertaken master's studies in intellectual property and competition law at the Catholic University. In 1998, he was awarded a Darwin Fellowship. Manuel has done research and acted as consultant for a wide range of institutions including the Andean Community, the Food and Agricultural Organization, the United Nations Development Programme, the United Nations Environment Programme, the International Development Bank, the International Development Research Centre and the International Union for the Conservation of Nature.

Pitambar Shrestha, project officer with LI-BIRD, has a bachelor's degree from Tribhuwan University, Nepal. He is an expert in the in-situ conservation of agricultural biodiversity and has over a decade's experience working with farmers and policymakers.

Fredy Sierra is an agronomist and manager of FIPAH. He is also an auxiliary professor, agronomy and rural development, CURLA at the National Autonomous University of Honduras, La Ceiba. He obtained his degree from CURLA in 1992.

Song Yiching is a social scientist with a special interest in rural development (especially working with women), farmer organizations and agricultural extension. She received a PhD in rural sociology and rural development studies from Wageningen University. Currently, she is a senior research scientist at the Center for Chinese Agricultural Policy, Chinese Academy of Sciences in Beijing, leading a long-term action research program to create synergies between the seed systems of farmers and the Chinese government. She has been leading the research in Guangxi since 1999.

B. Bahadur Tamang, project officer with LI-BIRD, has completed a master's degree in economics at Tribhuwan University, Nepal. For about six years, he has been involved in demonstrating and studying best practices in in-situ conservation of biodiversity, particularly in adding value to and marketing genetic resources.

Ronnie Vernooy is a rural development sociologist with a particular interest in agricultural biodiversity and natural resource management. He obtained his PhD from Wageningen Agricultural University, the Netherlands. He has more than 20 years of experience in managing and conducting participatory research in a number of countries, including Nicaragua, Cuba, Honduras, China, Nepal, Vietnam and Mongolia. He was a program officer at the International Development Research Centre, Canada, from 1992 until 2010. He has co-authored and co-edited several books and articles on biodiversity management and conservation, most recently with Song Yiching, *Seeds and synergies: innovating rural development in China*. He now works as an independent consultant.

Joseph Henry Vogel, PhD, is a professor of economics at the University of Puerto Rico–Rio Piedras who specializes in biodiversity conservation. An invited speaker at over 250 forums worldwide, he has authored *Genes for sale* and dozens of articles that explore an economic approach to access to genetic resources and intellectual property. He edited *The biodiversity cartel* and *The museum of bioprospecting, intellectual property, and the public domain: a place, a process, a philosophy*. He has served on the Ecuadorian delegation to the United Nations Conferences of the Parties (COP) of the Convention on Biological Diversity and his interests also extend to the UN Framework Convention on Climate Change. In Copenhagen at the COP 15, he launched *The economics of the Yasuní Initiative: climate change as if thermodynamics mattered*.

List of figures

List of tables

Abbreviations and acronyms

ABS	access and benefit sharing
BCDCs	biodiversity conservation and development committees
BMZ	Federal Ministry for Economic Cooperation and Development (Germany)
CBD	Convention on Biological Diversity
CBM	community-based biodiversity management
CBR	community biodiversity register
CBSP	community-based seed production
CCAP	Center for Chinese Agricultural Policy
CGIAR	Consultative Group on International Agricultural Research
CIAL	Comité de Investigación Agrícola Local (Honduras)
CIAT	Centro Internacional de Agricultura Tropical
CIDA	Canadian International Development Agency
CIP-UPWARD	Centro Internacional de la Papa—Users' Perspectives With Agricultural Research and Development
COP	Convention of the Parties
CSD	community seed bank
DICTA	Directorate of Science and Agricultural/Livestock Technology (Honduras)
DR-CAFTA	Dominican Republic–Central American Free Trade Agreement
EAP	Escuela Agrícola Panamericana (Honduras)
EU	European Union
FIPAH	Fundación para la Investigación Participativa con Agricultores de Honduras
GCSAR	General Commission for Scientific and Agricultural Research (Syria)
GMRI	Guangxi Maize Research Institute (China)
GRPI	Genetic Resources Policy Initiative
GRRI	Guangxi Rice Research Institute (China)
GTZ	Deutsche Gesellschaft für Technische Zusammenarbeit (Germany)
ICARDA	International Center for Agricultural Research in Dry Areas (Syria)

ICIMOD	International Centre for Integrated Mountain Development (Nepal)
IDRC	International Development Research Centre (Canada)
IGC	Intergovernmental Committee on Genetic Resources and Intellectual Property, Traditional Knowledge and Folklore
INCA	National Institute of Agriculture Sciences (Cuba)
INDECOPI	National Institution for the Defense of Competition and Intellectual Property (Peru)
ITPGRFA	International Treaty on Plant Genetic Resources for Food and Agriculture
JCC	Jordan Cooperative Corporation
LI-BIRD	Local Initiatives for Biodiversity, Research and Development (Nepal)
MATs	mutually agreed terms
MoAC	Ministry of Agriculture and Cooperatives (Nepal)
MoFSC	Ministry of Forests and Soil Conservation (Nepal)
NARC	National Agriculture Research Council (Nepal)
NBS	Nepal Biodiversity Strategy
NCARE	National Center for Agricultural Research and Extension (Jordan)
NDUS	new, distinct, uniform and stable
NGO	non-governmental organization
NTFP	non-timber forest product
NUS	neglected and under-used species
OAU	Organization for African Unity
PGR	plant genetic resources
PGRFA	plant genetic resources for food and agriculture
PIC	prior informed consent
PPB	participatory plant breeding
PPVFR	Protection of Plant Varieties and Farmers' Rights Act (India)
PSD	participatory seed dissemination
SAWTEE	South Asia Watch on Trade, Economics and Environment
SENASA	National Service of Plant and Animal Health (Honduras)
SMTA	standard material transfer agreement
SPDA	Sociedad Peruana de Derechos Ambientales
SUB	sustainable use of biodiversity
THPI	Tinjure Handmade Paper Industry
TRIPS	Trade Related Aspects of Intellectual Property Rights
UPOV	Union for the Protection of New Varieties of Plants
VDC	village development committee
WIPO	World Intellectual Property Organization
WTO	World Trade Organization

Websites

ABS Capacity Development Initiative for Africa	www.abs-africa.info
Bio Trade Initiative	www.biotrade.org
Convention on Biological Diversity	www.cbd.int
Food and Agriculture Organization of the United Nations	www.fao.org
Fundación para la Investigación Participativa con Agricultores de Honduras/ Foundation for Participatory Research with Farmers of Honduras	www.fipah.org
Fridtjof Nansen Institute	www.fni.no
Genetic Resources Policy Initiative	www.grpi.org
Global Crop Diversity Trust	www.croptrust.org
Global Environmental Facility	www.thegef.org
International Center for Agricultural Research in Dry Areas	www.icarda.org
International Centre for Integrated Mountain Development	www.icimod.org
International Development Research Centre	www.idrc.ca
International Institute for Sustainable Development	www.iisd.org
International Treaty for Plant Genetic Resources	www.treaty.org
International Union for the Conservation of Nature	www.iucn.org
Local Initiatives for Biodiversity Research and Development	www.libird.org
National Center for Agricultural Research and Extension	www.ncare.gov.jo
National Commission for the Prevention of Biopiracy	www.biopirateria.gob.pe
Participatory Action Research in China	www.parinchina.org
Potato Park Cusco	www.parquedelapapa.org

Sociedad Peruana de Derechos Ambientales	www.spda.org.pe
South Asia Watch on Trade, Economics and Environment	www.sawtee.org
United Nations Conference on Trade and Development	www.unctad.org
United Nations University	www.unu.edu
World Intellectual Property Organization	www.wipo.org
World Trade Organization	www.wto.org

Preface

A hand returned

Manuel Ruiz and Ronnie Vernooy

I have been involved in building access and benefit-sharing (ABS) frameworks since 1993. I participated in the design and development of the so-called Andean Regime, the first truly regional ABS policy framework, which was approved by the governments of Bolivia, Ecuador and Colombia in 1994. At the time, and even more so today, there was abundant evidence of inequity and asymmetries in terms of who benefited from flows of biodiversity and genetic components between the South and North. The "biopiracy phenomenon" certainly catalyzed my interest in working toward equity and fairness and safeguarding the social, cultural, legal and economic interests of our countries and their communities. Today, ending biopiracy is a key part of my work for SPDA (Sociedad Peruana de Derechos Ambientales or Peruvian Society for Environmental Rights, a non-governmental organization).

Unfortunately, I believe that regardless of all the claims and efforts of provider countries to strengthen national sovereignty and rights to genetic resources, it is highly unlikely that the key objectives of the Convention on Biological Diversity (CBD) will be met. Current national and regional ABS policies, frameworks and templates—including the Nagoya Protocol on Access and Benefit Sharing to Genetic Resources and the Fair and Equitable Sharing of Benefits adopted during the 10th Convention of the Parties to the CBD in Nagoya (COP 10) in October 2010—are, to a considerable extent, responsible for this state of affairs. The implementation of these frameworks and the Nagoya Protocol represent an enormous challenge.

I doubt, therefore, that fair access will result or that benefits will be shared equitably under the existing international regime and national and regional frameworks. I must admit that some of us are responsible for driving the existing ABS frameworks and templates in a certain direction without evaluating alternative approaches. In retrospect, I have no problem admitting we may have been wrong in our thinking.

If you critically review the current Nagoya Protocol on ABS, as well as national legislation already in place in a few countries and regions (Costa Rica and the Andean Community, for example), or even if you consider the much lauded Bonn Guidelines on ABS, which were adopted following COP 6 in 2002, you will find they all have the same conceptual orientation and share a similar structure and content. I believe that the inner logic of all these normative frameworks is colored by a disregard for new technological advances and how they affect everyday practices of using and maintaining biological diversity (basically, a misunderstanding of scientific processes). This is not good news, especially for provider countries or countries of origin, such as Peru, my country of birth, study and work.

Over the past few years, I have been paying more attention to this new technological paradigm and its impact on biodiversity resources and their governance. I find that Joe H. Vogel's suggestions (see, for example, Vogel 2010 and the epilogue of this book)—i.e. allowing for free flow, triggers at the commercial or patent stage, setting up a global fund, distributing benefits on a geographic scale—make good sense and might actually contribute to stimulating research and flows of resources. There are, of course, some hurdles to implementing this "proposal," but I see them as much less burdensome than the current ABS approaches, including those that were debated at COP 10. In my view, this "new" approach, which focuses on future monetary benefits, does not preclude mechanisms to foster technology transfer and non-monetary benefits.

I'd like to challenge the current political arena to consider this option carefully and make an effort to understand the science behind research and development in genetic resources.

Learning from past efforts and looking at how technology is advancing, I think redirecting the ABS process is absolutely necessary. A Turkish proverb tells us, "No matter how long you have gone down the wrong road, turn back." This advice increasingly resounds in the back of my mind when I think about the history of the international ABS process. However resistant politics has become to adopting novel ideas, I think there is no alternative.

Manuel Ruiz
Lima, Peru

Thank you, Manuel, for your frank and critical insights. My involvement with the international ABS negotiations is much shorter than yours, and I am not an experienced lawyer as you are. I came to hear and learn about ABS issues through my interest in farmers' livelihoods and the role seeds play in agricultural production—not only in countries of the South, but also in the North. My grandparents were farmers in Holland; I guess this is where my roots are.

Twenty-five years ago, when I first set out to research subsistence farmers' production practices as a student at Wageningen Agricultural University, I learned about a widely used local system of labor exchange called, in Spanish, *mano vuelta*. Literally, it means a hand returned. In practice, it is used to describe a variety of ways in which farmers exchange labor during a particular agricultural cycle or between two sequential cycles (sometimes three). Farmers make up for labor shortages by pooling resources when demand is high, when someone is sick or otherwise temporarily immobilized, or when an emergency situation appears.

Usually, labor is traded for labor, based on the number of days worked, but sometimes it is exchanged for a share in the harvest, e.g. a day of work equals a given number of bags of maize. Rarely does money come into play. Nicaraguan farmers using this system also apply it to the exchange of seeds. They give a bag of seeds to a neighbor or friend in need, expecting to receive a bag in return at some later date. Seeds for seeds: "yes, *mano vuelta*," they would say when I asked them what they called this system of reciprocity of access and sharing.

In 1992, the year the CBD was adopted, I contributed to the development of a new programming theme at the International Development Research Centre (IDRC) of Canada, in response to the CBD. The program was called, simply, Biodiversity. Later, from 1997 to 2005, I was a member of a more specific program named Sustainable Use of Biodiversity (SUB). This was a time when issues of "ownership" of genetic resources became more and more central to my work. Important ideas were captured in a number of publications supported by the SUB program, including a thought-provoking book by two leading researchers on questions concerning biological diversity. Darrell A. Posey and Graham Dutfield (1996) put forward strong arguments for safeguarding the biological resource rights of indigenous peoples around the world in light of ever-increasing attempts by transnational and national companies alike to patent "life."

Probably the largest and best-known initiative that dealt with ABS issues in those years was the Crucible Project. It brought together a diversity of actors from around the world to put forward options for international and national policies and laws that would ensure fair and equitable access to and sharing of benefits from biodiversity. *Mano vuelta* was raised to higher levels, one could say. Although not directly involved in the Crucible Group, I followed the lively debates with interest.

In 2003, I compiled the experience of ten years of IDRC support for participatory plant breeding in a book called *Seeds that give: participatory plant breeding*. Participatory plant breeding is an approach that pools the knowledge, labor, equipment, seeds and other resources of farmers and plant breeders (others, such as social scientists or extension agents, may also join forces) to improve crops and contribute to better rural livelihoods. In *Seeds that give*, there is no explicit mention of ABS per se, but the question of ownership of newly bred varieties (an important ABS question *and* farmers' right issue which will be addressed in many of the case studies in this book)—products of joint efforts—is raised in the concluding chapter which is entitled "A vision for the future." Allow me to quote from "Everything connects" on page 74:

> In this new environment where participatory plant breeding is accepted as the norm, it is only natural that local community-based agro-biodiversity conservation and improvement activities are connected to changes at the international and national policy levels. Thus there is opportunity for community input to global arrangements such as the CBD, the Food and Agricultural Organization's International Treaty on Plant Genetic Resources, and the World Trade Organization's agreement on the trade-related aspects of Intellectual Property Rights. In this way the global context supports the diversity of local efforts, and the local diversity informs and guides the global.
>
> Because it was a pioneer in this field, Nepal in 2012 is a leader and seen by many as an example to follow. Farmers' committees, made up of about equal numbers of women and men, now work closely with the formal sector in developing and evaluating new varieties, and in testing postharvest technology. Recognition

by the government of farmers' rights has not only brought new respect to rural communities, it has also raised the level of participation in community affairs and improved local economies. Biodiversity fairs are popular and well attended, and the winners at these fairs are invited to become members of local and regional variety-release committees.

Several of the initiatives highlighted in *Seeds that give*, including one in Nepal, have continued their groundbreaking work, complementing field experiments with action research on policy and the legal aspects of crop improvement and agricultural conservation. In Part 2 of this book, we offer a selection of how some of these initiatives are putting ABS into practice as well as informing national policymaking and even contributing directly to the policymaking process. Maybe my future vision of farmers as members of variety-release committees remains a dream, but considerable experience has been gained in terms of recognizing and rewarding farmers for their role as key custodians of genetic resources. Supported by researchers, farmers around the world show us the way. At the heart of their practices, we find *mano vuelta*! Perhaps, all of us can learn something from them.

Ronnie Vernooy
Ulaanbaatar, Mongolia

References

Posey, D.A. and Dutfield, G. (1996) *Beyond intellectual property: toward traditional resource rights for indigenous peoples and local communities.* International Development Research Centre, Ottawa, Canada. Available at: http://publicwebsite.idrc.ca/EN/Resources/Publications/Pages/IDRCBookDetails.aspx?PublicationID=413 (accessed 1 April 2011).

Vernooy, R. (2003) *Seeds that give: participatory plant breeding.* International Development Research Centre, Ottawa, Canada. Available at: http://publicwebsite.idrc.ca/EN/Resources/Publications/Pages/IDRCBookDetails.aspx?PublicationID=337 (accessed 1 April 2011).

Vogel, J.H. (ed.) (2010) *The Museum of Bioprospecting, Intellectual Property, and the Public Domain: A Place, A Process, A Philosophy.* Anthem Press, London.

Acknowledgments

Around the world in farmers' fields and homes, at local seed fairs, during community events and in meeting places of various kinds, new practices for conserving and improving agricultural genetic resources are being born. They are the result of arduous and often courageous efforts of small and large teams of experimenters and innovators who care about the source of all life: earth's biodiversity. In this book, we bring together seven of these teams. They share their motivation, ideas, research designs and results to ensure fair and equitable access to the agricultural genetic resources that sustain both local communities and nations at large. Many people have contributed to the case studies that aim to capture their experiences and the lessons they learned in a few pages.

The **authors of the Syria case study** thank their colleagues in the General Commission for Scientific and Agricultural Research of the Ministry of Agriculture and Agrarian reform of the Syrian Arab Republic for their help over the years; Mr Michael Michael, Mr Adonis Kourieh, Mr Henry Pashayani, Mr Raafat Azzo, Mr Adnan Ayyan, Mr George Kashour, Mr Mahmoud Hamzeh and Ms Hanadi Al-Naji for their technical assistance; and several hundred farmers for sharing their knowledge and time with us, thus making this work possible.

Adnan Al-Yassin, the author of the Jordan case study, is very grateful to the farmers for their interactions and interest in adopting newly developed varieties. Their innovative ideas and enthusiasm were the driving force behind the work. He also thanks the Jordanian researchers in the project, in particular Ms Siham Al-Louzi, for her technical assistance in preparing the manuscript; and Ronnie Vernooy and Salvatore Ceccarelli for providing the opportunity to the farmers in Jordan and this chance to address their rights, access to and benefits from their genetic resources.

The **Honduras team** would like to thank their farmer research partners in the CIALs, who have been working with them for many years as collaborators in plant breeding and research. It is their enthusiasm for improving their seeds and for learning along with the team that motivates its work. However, motivation alone is not enough. Without the support of generous donors, none of the breeding work would have been possible. The team recognizes the financial support of USC-Canada, the Canadian International Development Agency, the Development

Fund of Norway, the International Development Research Centre of Canada (IDRC) and the Participatory Research and Gender Analysis small-grants program for participatory plant breeding within the Consultative Group on International Agricultural Research. These donors have provided more than financial assistance; they have provided moral support and guidance over an extended period. Finally, the team recognizes the technical support provided by the Pan American Agricultural School, Zamorano. The ideas contained in this article and any errors, however, are the responsibility of the authors alone.

The **China team** would like to thank the hundreds of women and men farmers in the southwest part of the country who have inspired their efforts, the courageous plant breeders from the Guangxi Maize Research Institute in Nanning and the Institute of Crop Science of Chinese Academy of Agricultural Science in Beijing who dared to walk unknown paths, the management team and other colleagues from the Center for Chinese Agricultural Policy who have supported them all these years and program officers at the Ford Foundation and IDRC, Canada, who had faith in their ideas and dreams.

The **authors of the Cuban case study** are grateful to Liss, Richard and Lilly from the Canadian International Development Agency and Olivier, Rodolfo and Herbert from the Swiss Agency for Development and Cooperation for their advice and funding. Special thanks go to Jose Roberto from the National Institute of Agricultural Sciences for being patient, enthusiastic and friendly in discussing the ideas and implementing the program. Thanks also go to Eva from ACSUR Las Segovias, Richard and Juergen of the German organization, Agro-Action, as well as to Susie and Beatriz of USC Canada who believed in and supported the dissemination efforts. Enormous thanks to all researchers, technicians and farmers of the team: they are the real concept makers and shakers of participatory seed diffusion. Thanks to Nathaniel and Julia for reviewing the Cuba chapter and for all the discussions.

The **Nepalese authors** are very grateful to the management teams of Local Initiatives for Biodiversity, Research and Development (LI-BIRD) and South Asia Watch on Trade, Economics and Environment (SAWTEE). They are indebted to the fieldworkers, Mr Shreeram Subedi, Ms Laxmi Rai, Mr Birendra Chaudhari, Mr Durga N. Shrestha and Mr Pashupati Khaniya, and to the communities in Begnash, Kachorwa, Jogimara, Bachhayauli and Tamaphok for their contribution in collection and making sense of the information. The authors would also like to acknowledge the contribution of other professional staff at LI-BIRD and SAWTEE. Last, but not least, they acknowledge the support of IDRC, Canada, for action research in Nepal carried out through the "Fair access and benefit sharing of genetic resources: national policy development in China, Jordan, Nepal, Peru" project.

The **editors** thank the seven teams for their inspiring contributions that are the heart of this book. They also express gratitude to Bert Visser and Niels Louwaars in the Netherlands and to colleagues (current and former) at Sociedad Peruana de Derechos Ambientales and at IDRC, especially Wardie Leppan, Brian Davy and

Jean Woo, who shared their passion and knowledge about access and benefit issues with us from very early on. Special thanks go to Tim Hardwick and other Earthscan colleagues and to Isabel Bortagaray and Bill Carman at IDRC who skillfully guided us through the book production process. Sandra Garland harmonized the various voices artfully.

Part 1

The global debate

1 Introduction

Widening the horizon

Ronnie Vernooy and Manuel Ruiz

Everywhere, local practices in biodiversity conservation, crop improvement and natural resource management are under stress. In some countries, communities and governments have to deal with "biopiracy" (Robinson 2010). Although there is no universal definition for the concept of biopiracy, a useful working meaning may be, "illegally accessing and using genetic resources and related traditional knowledge, either through direct appropriation or indirectly through the use of intellectual property, especially patents." This is based on Law 28216, which created the National Biopiracy Prevention Commission in Peru (more about this commission in Chapter 5).

Existing laws and mechanisms, such as intellectual property rights, are unable to protect indigenous and traditional knowledge and are inadequate when it comes to collaboratively developed innovations (such as varieties resulting from participatory plant breeding (PPB)), because, it is argued, they protect individual as opposed to collective rights. In PPB, farmers, researchers and others join forces to improve existing varieties or develop new ones, based on shared knowledge and resources. The improved or new varieties have multiple creators, whose efforts often build on the field experiments of previous generations. To whom do these varieties belong? Rights of access, use and sharing of benefits no longer reside with a professional plant breeder, and new definitions of these terms are required (Vernooy 2003).

A number of national and international policy processes are underway to allow for the development of *sui generis* systems—in simple terms, locally grown and appropriate systems—to protect local natural and genetic resources and related knowledge about their management, use and maintenance. The best known are the longstanding negotiations of the International Regime on Access to Genetic Resources and Benefit Sharing under the Convention on Biological Diversity (CBD (United Nations 1992)) and those of the International Regime for the Protection of Traditional Knowledge under the World Intellectual Property Organization (WIPO).

Adopted in 2001, the International Treaty on Plant Genetic Resources for Food and Agriculture (ITPGRFA), while recognizing national sovereign rights over plant genetic resources, represents a multilateral system for facilitated access to a limited number of agricultural crops for the fair and equitable sharing of the benefits arising from their use. However, so far and despite the recent agreement on an International Regime on Access and Benefit Sharing (the Nagoya Protocol; more about this in Chapter 2), progress has been painful and slow, and few concrete, workable results have been produced so far. Effective implementation of the Nagoya Protocol by national governments now looms on the horizon. Harmonization of the terms of the Nagoya Protocol with the ITPGRFA will be another important task.

One challenge has been to broaden the policy and legal debates beyond the sphere of policymakers, lawyers and other experts by including knowledge-holders themselves, i.e. farmers, herders and fishers—the day-to-day custodians of biodiversity—in the definition of questions and in the formulation, testing and assessment of answers. Another challenge is the fragmentation and confusion among those involved in national and international debates on access and benefit sharing (ABS). Disputes abound, reflecting different and what are perceived to be opposing interests.

This book presents promising examples of feasible and fair local ABS practices, as inputs for the development of innovative policy and legal alternatives at international, national and local levels. The examples are grounded in the practices of local and indigenous farming communities and linked to new partnership configurations of multiple stakeholders interested in supporting these communities. The results fill an important gap in current scientific and policy work and complement a number of interesting studies that have been completed and published recently (e.g. Kamau and Winter 2009, Richerzhagen 2010 and, focusing on China, Song and Vernooy 2010). For earlier discussions of ABS issues, see, for example, Crucible Group 1994, Poscy and Dutfield 1996, Lesser 1998, Bass and Ruiz Muller 1999, Crucible Group 2 2000, 2001, ten Kate and Laird 2002, Carrisoza *et al.* 2004; for a more general discussion in relation to food, see, for example, Tansey and Rajotte 2008.

The effective and fair implementation of mechanisms supported by appropriate policies and laws will ultimately be the most important assessment factor of any ABS regime. Local-level learning examples are key inputs for the development of national and international agreements and, most of all, for their effective implementation.

Access and benefit sharing and the Convention on Biological Diversity

Not long ago, the concepts of genetic resources, genetic diversity and biodiversity were confined to the closed realm of scientific discourse, with only marginal day-to-day use by non-scientists and citizens in general. Political, economic and legal analysis of these concepts was also very limited.

This situation changed with the CBD, which was adopted on 22 May 1992, signed during the United Nations Conference on the Environment and Development on 5 June 1992 and took effect on 29 December 1993. The CBD's objectives are to ensure the conservation and sustainable use of biodiversity, as well as the sharing of benefits from access to and use of genetic resources. Today, these concepts are more and more part of everyday policy-, social-, development- and environment-related discussions. Globally, a broad set of social actors is more conscious and aware of the critical importance of genetic resources and components of biodiversity, in general, in ensuring future viability of the planet and humanity. Today, most people are aware that plant and animal species are under pressure in many areas around the globe, even though they may not be directly engaged in activities to halt this trend.

The social, economic and environmental importance of biodiversity cannot be overlooked. Whether in terms of molecules for pharmaceutical research, food to support a hungry world or farmers in arid regions maintaining drought-resistant varieties of grains that may hold the key to adapting crops to climate change, biodiversity is the only hope if we are to achieve sustainable rural and urban livelihoods. Seeds, for example, are an illustrative and hugely important component of biodiversity—although they are not always valued as such.

Currently, presidents and local governors, bankers and entrepreneurs, public officials and indigenous peoples' representatives all use the concept of biodiversity and acknowledge its significance. The last decade or so has seen considerable global progress in efforts to raise awareness of biodiversity conservation. This has gradually resulted in policy and legal initiatives that have led to national biodiversity laws, biodiversity strategies and action plans, a vast array of biodiversity conservation programs and projects, and continued calls for concerted global action to conserve biodiversity (some of these initiatives are reviewed in Chapter 3).

In this regard, the CBD has played a pivotal role in triggering national (and global) action toward realization of its three main objectives: conservation of biodiversity, sustainable use of its components and fair and equitable sharing of the benefits derived from access to and use of genetic resources.

This third objective and related principles have received the most attention over the past decade, including recommendations on how to achieve ABS. International discussions on ABS mirror those on a series of larger complex political and socioeconomic problems, which in many cases are the expression of tensions and sometimes conflicts between countries around the world that sometimes join forces in blocs. In brief, the focus is on such questions as: Access to what, exactly? Access for what purposes? What are the envisioned benefits? Whose benefits are they? What is fair and equitable sharing? How can this kind of sharing be made to work?

For more than a decade, debates on benefit sharing have been mainly limited to situations involving genetic resources, where biotechnology is applied and intellectual property rights are invoked to protect innovations resulting from research and development efforts to use biodiversity. Indeed, the original rationale for the inclusion of this third CBD objective was to seek a balance between the interests of biodiversity-rich but technologically limited countries (historically net providers of genetic resources) and biodiversity-poor, developed, but technologically advanced countries. ABS debates are another reflection of the North–South tension that has polarized the world for decades. Establishing national and international ABS policies, laws and strategies is one way to resolve these tensions.

The justification for putting ABS issues on the table (in direct association with genetic resources, traditional knowledge, biotechnology and intellectual property rights) can be traced back to the late 1970s when a group of concerned scientists and activists from around the world began to question and reflect on the policy, legal and economic implications surrounding the flow of genetic resources and biodiversity in general (in the context of the rapid loss of biodiversity in many places around the world). The transformation of these resources through biotechnology and their subsequent appropriation through patents rapidly divided the world into two groups: the industrialized North in search of genetic resources and the biodiversity-rich South (for a detailed, historical overview on the political and social implications of the early debates, see, for example, Pistorius 1997; for a personal perspective, see Mooney 2011).

Although the development of ABS policies, laws and strategies is one way to deal with these tensions, the road to resolution has been a rocky one. The main reason for this treacherous road, we argue, has been the prevalence of a rather narrow view of ABS, i.e. access to and use of *genetic* resources. This limited view has been characterized by a very legalistic approach to the CBD. However, more recently, experts and a series of actors involved in the conservation, management and sustainable use of biodiversity, in general, have started to look at how benefit sharing is expressed in contexts other than genetic resources, including conservation and sustainable use of biological resources, scientific research, plant breeding and seed conservation, especially when they involve activities undertaken by indigenous and local communities and farmers at the local level.

These actors are taking a broader, more holistic and dynamic view of ABS. They have also drawn attention to the fundamental roles that the custodians of biodiversity play in safeguarding life on earth, roles that both are shaped by and that shape the larger political economy. This has led to strong advocacy of the recognition of farmers' rights to acknowledge these vital important roles. The ITPGRFA explicitly recognizes farmers' rights and encourages contracting parties, subject to their national legislation, to take measures to protect and promote them. However, how to implement and ensure farmers' rights remains a challenge.

This book

This publication presents concrete examples and an analysis of how ABS is conceptualized and practiced in contexts other than the narrowly focused view on access to genetic resources. It does not "throw away" ABS altogether, but, rather, broadens the issue to situate genetic resources in the context of the dynamic evolution of local (rural) livelihoods. This is important, because, increasingly, practical cases are being sought to allow better understanding of how benefit sharing can actually work—turning on its head the conventional wisdom that policies and laws come first to guide human behavior. Instead, this book presents pioneering examples of ABS in practice, in areas such as seed conservation, biopiracy prevention and PPB, from which more general guidelines are emerging for the regulation of the use and conservation of biodiversity. The cases also serve as concrete examples of ways to implement existing policies or policy frameworks (or components of them), including the ITPGRFA and farmers' rights, through national policies, laws and other regulatory frameworks.

The need for this broader look at ABS is especially acute in international discussions, as a way to better inform debates and allow policymakers to identify the wide and varied conditions and situations where ABS occurs or could occur. As such, we hope that these grounded examples and the analysis derived from them will allow for the generation of more effective policies and laws that regulate ABS and, ultimately, contribute to the larger goals of conservation and sustainable livelihoods.

This book is divided into two parts and three chapters (Part 1) plus nine chapters (Part 2). Part 1 offers an overview of the broad global policy debates on ABS. The brief introduction to the key issues and problems in this chapter is followed, in Chapter 2, by an analysis of what ABS means in conceptual and legal terms. This chapter builds on the historic evolution of the concept of "benefit sharing," then analyzes how it has been incorporated into international (i.e. CBD, ITPGRFA) and national legal instruments. Chapter 3 provides a summary of some of the better-known examples of ABS projects and initiatives worldwide—their goals, participants and impacts. It is based on a review of literature, but also draws on the personal experience of the authors through their direct involvement in some of the initiatives described.

Part 2 is the core of this publication. Chapter 4 introduces a series of case studies, which are detailed in Chapters 5–11. The studies—in Peru, Syria, Jordan, Honduras, China, Cuba and Nepal—all deal with conservation and the sustainable use of biodiversity (and genetic resources) as part of farmers' livelihood strategies. From PPB to in-situ conservation efforts and from preservation of traditional knowledge to commercialization and utilization of biodiversity components, these case studies offer important lessons regarding not only how benefit sharing expresses itself in practice, but also how it is understood and perceived by communities and actors directly involved in in-situ and ex-situ conservation, research, development, seed exchange and protection of traditional knowledge. In addition, some of the cases explicitly address farmers' rights. These chapters are written by researchers from the various countries together with their partners, local or international. They are based on long-term field research and involvement in advocacy and policymaking efforts.

All case studies follow a common outline: the criteria that local actors, policymakers and researchers use to value genetic resources and biodiversity; how projects and activities determine the areas, sites and ecosystems affected; how traditional knowledge is researched, processed and systematized; how value is added to local knowledge and innovation; how and under what protocols and guidelines projects and activities are undertaken; and how results are communicated and exchanged among various actors. As part of this continuum of activities, benefits flow in different ways among various actors—sometimes obviously and explicitly; sometimes in indirect ways; in monetary terms; through exchange of knowledge, seeds, breeding materials and ecosystem management information; and in terms of enhanced participation by local communities in research, policy, legislative and regulatory processes. The cases are introduced in more detail in Part 2.

Finally, Chapter 12 offers a comparative overview of the cases, tries to answer key questions concerning ABS in practice and concludes with some final comments that may inform policy and legal processes addressing benefit sharing at the national, regional and international levels, including the CBD and the ITPGRFA. We hope these comments will contribute to ongoing efforts by individual experts, organizations and local communities around the world to develop sound, practical and effective policies and legislation, to implement and assess them, and to adapt them when necessary.

The book closes with an Epilogue entitled "Architecture by committee and the conceptual integrity of the Nagoya Protocol," written by Joseph H. Vogel. His spirited and critical contribution brings the reader back to the bigger picture of central ABS issues and questions as debated internationally and nationally.

May this book inspire current and future generations of custodians to not give up, but to continue innovating.

References

Bass, S.P. and Ruiz Muller, M. (eds.) (1999) *Protecting biodiversity: national laws regulating access to genetic resources in the Americas.* International Development Research Centre, Ottawa.

Carrisoza, S., Brush, S.B., Wright, B.D. and McGuirre, P.E. (2004) *Accessing biodiversity and sharing the benefits: lessons from implementation of the CBD.* International Union for the Conservation of Nature, Gland, Switzerland.

Crucible Group (1994) *People, plants, and patents.* International Development Research Centre, Ottawa. Available at: http://publicwebsite.idrc.ca/EN/Resources/Publications/Pages/IDRCBookDetails.aspx?PublicationID=251 (accessed 1 April 2011).

Crucible Group 2 (2000) Policy options for genetic resources: *People, plants, and patents* revisited. In *Seeding solutions*, vol. 1. International Development Research Centre, Ottawa, Canada; International Plant Genetic Resources Institute, Rome, Italy; Dag Hammarskjöld Foundation, Uppsala, Sweden. Available at: http://publicwebsite.idrc.ca/EN/Resources/Publications/Pages/IDRCBookDetails.aspx?PublicationID=246 (accessed 1 April 2011).

—— (2001) Options for national laws governing control over genetic resources and biological innovations. In *Seeding solutions*, vol. 2. International Development Research Centre, Ottawa, Canada; International Plant Genetic Resources Institute, Rome, Italy; Dag Hammarskjöld Foundation, Uppsala, Sweden. Available at: http://publicwebsite.idrc.ca/EN/Resources/Publications/Pages/IDRCBookDetails.aspx?PublicationID=248 (accessed 1 April 2011).

Kamau, E.C. and Winter, G. (eds.) (2009) *Genetic resources, traditional knowledge and the law: solutions for access and benefit sharing.* Earthscan, London and Sterling.

Lesser, W. (1998) *Sustainable use of genetic resources under the CBD: exploring access and benefit sharing issues.* CABI, Wallingford, UK.

Mooney, P. (2011) The hundred year (or so) seed war: seeds, sovereignty and civil society. A historical perspective on the evolution of "the war of the seed." In C. Frison, F. López, J.T. Esquinas (eds.) *Plant genetic resources and food security: stakeholder perspectives on the International Treaty on Plant Genetic Resources for Food and Agriculture.* Earthscan, London and Sterling (forthcoming).

Pistorius, R. (1997) *Scientists, plants and politics: a history of the plant genetic resources movement.* International Plant Genetic Resources Institute, Rome, Italy.

Posey, D.A. and Dutfield, G. (1996) *Beyond intellectual property: toward traditional resource rights for indigenous peoples and local communities.* International Development Research Centre, Ottawa, Canada. Available at: http://publicwebsite.idrc.ca/EN/Resources/Publications/Pages/IDRCBookDetails.aspx?PublicationID=413 (accessed 1 April 2011).

Richerzhagen, C. (2010) *Protecting biological diversity: effectiveness of access and benefit sharing regimes.* Routledge, London.

Robinson, D.F. (2010) *Confronting biopiracy: challenges, cases and international debates.* Earthscan, London and Sterling.

Song, Yiching and Vernooy, R. (eds.) (2010) *Seeds and synergies: innovating rural development in China.* Practical Action Publishing, Bourton on Dunsmore, and International Development Research Centre, Ottawa.

Tansey, G. and Rajotte, T. (eds.) (2008) *The future control of food: a guide to international negotiations and rules on intellectual property, biodiversity and food security.* Earthscan, London and Sterling, the Quaker International Affairs Program, and International Development Research Centre, Ottawa.

ten Kate, K. and Laird, S.A. (2002) *The commercial use of biodiversity: access to genetic resources and benefit sharing.* Earthscan, London and Sterling.

United Nations (1992) Convention on biodiversity. United Nations, New York, NY, USA. Available at: www.cbd.int/doc/legal/cbd-en.pdf (accessed 1 April 2011).

Vernooy, R. (2003) *Seeds that give: participatory plant breeding.* International Development Research Centre, Ottawa, Canada. Available at: www.idrc.ca/en/ev-30294-201-1-DO_TOPIC.html (accessed 1 April 2011).

2 The policy and legal context for access and benefit sharing

Manuel Ruiz and Ronnie Vernooy

Principles of access and benefit sharing

The notion of ABS is relatively new. It is most notably expressed as a principle in various sections of the CBD (UN 1992). In that context, ABS is inextricably linked to genetic resources and the related traditional knowledge of indigenous peoples and communities. According to the CBD, the action of *accessing* genetic resources, on which, ultimately, everyone on earth depends for survival (FAO 2010) triggers a requirement for sharing, fairly and equitably, all the benefits derived from their use and from the related intellectual efforts of indigenous peoples and communities to maintain them over time. In the spirit of the CBD, one cannot freely access genetic resources without considering how they will be used, by whom and under what conditions.

But how will the noble principles of fairness and equity be realized in practical terms? In the international arenas concerned with the notion of ABS, this question has been discussed continuously; outside these arenas, action-oriented researchers, NGOs, local government staff, together with farmers and other local people have been experimenting with this concept in practical ways. There is no blueprint for this work; it is done on a case-by-case basis, starting from local realities. The one element on which there is firm consensus is that benefits derived from the use of genetic material can be both monetary and non-monetary (Table 2.1). Full and clear consensus on other elements remains a work in progress, notwithstanding the recent adoption at COP 10 of the very general International Protocol on Access to Genetic Resources and the Fair and Equitable Sharing of Benefits Derived from their Use (Nagoya Protocol), which was heralded by several United Nations' staff as "historic" (UN News Centre 2010).

The Convention on Biological Diversity

In its preamble, the CBD first refers to "benefit sharing" in the context of indigenous peoples (UN 1992). It expressly recognizes "the desirability of sharing equitably benefits arising from the use of traditional knowledge, innovations and practices, relevant to the conservation of biological diversity and the sustainable use of its components." The obligation to share benefits is triggered when

Table 2.1 Potential monetary and non-monetary benefits derived from the use of genetic resources

Monetary	Non-monetary
• Up-front fees (cost of samples, shipping, handling, etc.) • Milestone payments at different stages of the research and development process • Royalties (after commercialization)	• Participation of providing-country scientists in research • Transfer of technology and equipment • Staff exchanges • Training • Acknowledgments in publications • Sharing of results of research • Voucher specimens • Joint patents or intellectual property rights • Social recognition • Creation of local or national biodiversity or rural development fund

Source: Adapted and extended from ten Kate and Laird 2002.

indigenous intellectual efforts are effectively accessed and used in some way. The preamble does not stipulate obligations. However, it inspires the content of the CBD and recognizes specific principles relevant to contracting parties and other actors in the interpretation and implementation of its substantial provisions.

This principle must be read in conjunction with Article 8(j), In situ conservation, which determines that CBD contracting parties should "respect, preserve and maintain knowledge, innovations and practices of indigenous and local communities embodying traditional lifestyles relevant for the conservation and sustainable use of biological diversity and promote their wider application with the approval and involvement of the holders of such knowledge, innovations and practices and encourage the equitable sharing of the benefits arising from the utilization of such knowledge, innovations and practices." Under this article, benefit sharing is conditional on the use of knowledge, innovations and practices.

The CBD includes a series of provisions throughout its text, calling for benefit sharing in the specific context of access to genetic resources. Article 1 defines the CBD's objectives, which include conservation of biodiversity, the sustainable use of its components and "the fair and equitable sharing of the benefits arising out of the utilisation of genetic resources, including by appropriate access to genetic resources and by appropriate access to relevant technologies."

Article 15, Access to genetic resources, addresses and refines the rules and principles applicable to access to and use of genetic resources. It highlights the recognition of the sovereignty of states over natural resources (and, therefore, their ability to regulate how and under what conditions genetic resources can be used) and refers to mutually agreed terms and prior informed consent, both expressed mainly through contracts. Such contracts or agreements are between providers (the state, individuals, ex-situ centers, communities) and users of resources (researchers, companies, universities, etc.). Thus, to put it simply, the

CBD takes genetic resources out of the commons through a particular form of privatizing them under state protection.

Article 15.7 specifically states that each contracting party shall take legislative, administrative or policy measures "with the aim of sharing in a fair and equitable way the results of research and development and the benefits arising from the commercial and other utilization of genetic resources with the Contracting Party providing such resources. Such sharing shall be upon mutually agreed terms."

Finally, Article 19.2, Handling of biotechnology and distribution of its benefits, stipulates that the contracting parties take measures "to promote and advance priority access on a fair and equitable basis by Contracting Parties, especially developing countries, to the results and benefits arising from biotechnologies based upon genetic resources provided by those Contracting Parties. Such access shall be upon mutually agreed terms."

In terms of Articles 1, 15.7 and 19.2, benefit sharing focuses primarily on access to and use of genetic resources and the goods and services produced from them. These articles describe with some specificity the types of benefits that may be involved and require sharing: technologies (and their transfer), research and development results arising from the use of biotechnology, commercial (monetary) benefits, among others which may arise. At the same time, it is no minor issue that these articles refer to genetic resources, not biodiversity or biodiversity components in general. This widely accepted interpretation of these CBD provisions has had the effect of narrowing discussions and implementation efforts to the development of ABS policies, laws and regulations on genetic resources and related issues—traditional knowledge protection, intellectual property and biotechnology. As part of the Nagoya Protocol on ABS, benefit sharing should also apply to access and use of "derivatives" (i.e. oils, resins, natural compounds).

The Bonn Guidelines on Access to Genetic Resources, which were adopted at the Sixth Conference of the Parties of the CBD (Decision VI/24), held in The Hague, the Netherlands in 2002, also focus on genetic resources and, in general, on a narrow set of issues associated with them including benefit sharing, traditional knowledge, intellectual property and biotechnology.

The Nagoya Protocol

After dramatic last-minute negotiations, the International Protocol on Access to Genetic Resources and the Fair and Equitable Sharing of Benefits Derived from their Use was adopted during the Tenth Conference of the Parties in Nagoya, Japan, October 2010 (COP 10).

This new international, binding instrument establishes the rules and principles that govern how and under what conditions genetic resources can be obtained and used. These conditions are based on prior informed consent and terms agreed to by countries of origin and users. The Nagoya Protocol seeks to ensure more equitable and fair terms governing access to, use of and exchange of genetic resources in in-situ conditions, especially in countries in the South.

The key advances of the Nagoya Protocol are in the area of compliance, traditional knowledge, basic research exemptions, the possibility of developing multilateral rules for use of genetic resources that are shared among countries and the recognition of a certificate of compliance to accompany the flow of resources during the research and development process.

In terms of compliance, countries are now obliged not only to adopt measures to facilitate access to their genetic resources, but also to ensure that, when using genetic resources from foreign jurisdictions, they adopt measures to safeguard the interests of providing countries. So-called "user measures" are seen as the only way to overcome difficulties regarding, for example, compliance of contractual conditions or verifying flows of resources. These may also include adjustments to legislation to include mandatory disclosure of origin and legal provenance when intellectual property rights are sought.

The protocol also provides for respect for and protection of traditional knowledge as an essential part of biodiversity and its components. It proposes "biocultural protocols" and agreements (among other instruments) that will help define access to and use of traditional knowledge by third non-indigenous parties for scientific research and, eventually, commercially oriented research. Negotiations on protection of traditional knowledge are part of a broader international agenda, specifically of the World Intellectual Property Organization (WIPO) and its Intergovernmental Committee on Genetic Resources and Intellectual Property, Traditional Knowledge and Folklore (IGC). In 2009, the IGC received a mandate from WIPO's General Assembly to initiate negotiations on an international regime for the protection of traditional knowledge.

Non-commercial research (e.g. taxonomic investigation) is considered an exceptional case of ABS and, therefore, a simplified access procedure and contract is required to ensure continued and non-impeded basic biodiversity research. This is viewed as a major success by the scientific community.

The protocol also recognizes the possibility of adoption of future multilateral rules and principles governing ABS of shared genetic resources. Although included as an exceptional situation, this is potentially the most frequent scenario, given how biodiversity and genetic resources are shared nowadays across borders and jurisdictions, especially if we consider the mega-regions of diversity in Asia, Africa and Central and South America. When genetic resources are understood as coded genetic information, the notion of shared resources and information becomes even more apparent.

It has been said that the focus of the Nagoya Protocol is limited to genetic resources as such, mainly those with potential use in the pharmaceutical, cosmetics, bioremediation and biotechnological sectors. Although derivatives (biochemical compounds that result from natural metabolism of biological and genetic resources) are also included within the scope of the protocol, its full application to the wide range of products considered derivatives is still a matter for interpretation. Interpretation of "utilization of genetic resources" to include research into and development of the genetic or biochemical composition of genetic resources offers one way to extend the scope of the protocol to cover derivatives.

Finally, the multilateral flows of genetic resources for food and agriculture and the specific ABS provisions that guide them, covered by the ITPGRFA (FAO 2009), have been excluded from the protocol. This treaty was adopted in 2001 by the 31st session of the FAO conference and it entered into force on 29 June 2004. It established a special mechanism: the Multilateral System on Access and Benefit Sharing. Interdependence of all countries on foreign or imported seeds for agriculture and the critical importance of seed in ensuring food security led negotiators to agree on a multilateral mechanism by which resources are shared through a standard material transfer agreement and benefits are shared by all parties to the FAO treaty (for a detailed, multi-stakeholder perspective on the historical process that led to the ITPGRFA, see Frison *et al.* 2011).

The International Treaty on Plant Genetic Resources for Food and Agriculture

The objectives of the ITPGRFA closely follow those of the CBD. They are "the conservation and sustainable use of plant genetic resources for food and agriculture and the fair and equitable sharing of the benefits arising out of their use, in harmony with the Convention on Biological Diversity, for sustainable agriculture and food security" (Article 1.1). Indeed, the treaty states that these objectives will be attained through a close link to "the Food and Agriculture Organization of the United Nations and to the Convention on Biological Diversity" (Article 1.2). The treaty encourages contracting parties (i.e. national governments) to "develop and maintain appropriate policy and legal measures that promote the sustainable use of plant genetic resources for food and agriculture" (Article 6.1). Suggested measures include: fair policies that promote diverse farming systems; research support that enhances biological diversity (intra- and inter-specific variation); broadening of the genetic base of crops in situ and ex situ; support for the use of local crops, varieties and underutilized species; promotion of PPB; the creation of stronger links to plant breeding and agricultural development; and review/ adjustment of breeding strategies and regulations concerning variety release and seed distribution (Article 6.2).

As mentioned in Chapter 1, the treaty pays special attention to farmers' rights, stating that their realization rests with national governments (Article 9). Three rights are mentioned: protection of traditional knowledge relevant to plant genetic resources for food and agriculture (PGRFA); the right to participate equitably in benefits arising from the use of PGRFA; and the right to participate in making decisions, at the national level, on matters related to the conservation and sustainable use of PGRFA (Article 9.2). A final sub-article (9.3) adds that "nothing in this Article [9] shall be interpreted to limit any rights that farmers have to save, use, exchange and sell farm-saved seed/propagating material, subject to national law and as appropriate."

However, the crux of the treaty concerns the multilateral flows of a defined number of PGRFA: 20 major food crops, 15 legume forage crops, 12 grass forage crops and 2 other forage crops (for details, see Annex I to the treaty). Such flows will be encouraged through the signing of a standard material transfer agreement (SMTA), adopted in 2006, the general content of which is detailed in Article 12. Benefit-sharing provisions are detailed in Article 13. Proposed benefit-sharing mechanisms are the exchange of information, access to and transfer of technology, capacity-building and the sharing of the benefits arising from commercialization. Parties to the ITPGRFA are entitled to benefit sharing in as much as they are signatories and participate in the multilateral system of ABS (Part IV of the treaty). Benefits do not result directly from each specific SMTA, but from participation and commitment to the exchange and flows of seeds within the system.

Especially important in the ITPGRFA are its provisions on monetary benefits and intellectual property. Article 13.2.d and the SMTA determine that if monetary benefits arise from commercial use of seeds covered by the multilateral system (i.e. sales), a percentage of these benefits (1.1% of sales of products minus 30% of this) will be destined for the FAO Trust Fund (Annex 2 of the SMTA). Concerning intellectual property, seeds should be available with no restrictions for research and development. However, the treaty recognizes that plant breeders' rights and, in some cases, patents may be invoked. Article 12.3.d specifies that "Recipients shall not claim intellectual property or other rights that limit the facilitated access to the plant genetic resources for food and agriculture, or their genetic parts or components, *in the form received* from the Multilateral System." The general interpretation of this article is that only intellectual property rights that limit access cannot be claimed (i.e. patents cannot be claimed, but plant breeders can claim rights). Also, intellectual property rights over genetic resources and their components in the form received from the system are forbidden; however, if they are modified or developed, they may be subject to protection. Finally, the ITPGRFA is also part of a broad set of FAO-related policy (and legal) instruments that address conservation and the sustainable use of agricultural biodiversity, including the FAO Global Plan of Action for the Conservation and Sustainable Use of Plant Genetic Resources for Food and Agriculture (1996), the FAO Global System on Plant Genetic Resources for Food and Agriculture (1983) and the Leipzig Declaration on the Conservation and Sustainable Use of Genetic Resources for Food and Agriculture (1996), among others.

Access and benefit-sharing policies and laws

A rapid overview of existing ABS policies and regulatory frameworks in the Andean Community (Bolivia, Colombia, Ecuador and Peru), Brazil, Costa Rica, the African Union (comprising 53 countries), the Philippines, Panama and a few other countries around the world suggests that there are striking similarities in their content and format. Furthermore, all focus on physical access to genetic resources, establish some legal conditions for their transformation, protect traditional knowledge (if it is involved), limit the restrictions that intellectual property rights may place on derived innovations and establish mechanisms for participation in monetary and non-monetary benefits (Carrizosa *et al.* 2004). The Nagoya Protocol clearly builds on these policies and frameworks, which, pioneering as they have been, have proved hard to implement in practical ways over the past 15 years or so.

Furthermore, the narrow definition of "genetic resources" has limited the scope and coverage of these laws and policies. According to the CBD, genetic resources are "genetic material of actual or potential value;" genetic material is defined as "any material of plant, animal, microbial or other origin containing functional units of heredity." However, in the process of developing national and regional ABS regulatory frameworks, their scope has expanded to cover derivatives or derived products, which may include (depending on each case) natural oils, resins, extracts of all types, bark, mixtures, etc. Even though the Nagoya Protocol has arguably included these, how to deal with derivatives and the far-reaching implications surrounding their use (think of the cosmetics industry, nutraceuticals, botanical medicines, natural products, etc.) remains a point of contention and will need to be further interpreted as the protocol is implemented.

In terms of "benefit sharing," a few countries are using policy and legal instruments to extend this notion to other areas. For example, in case studies for Bolivia, Cuba and Ecuador, references are made to the need for equity and fairness in the sharing of benefits derived from the use of biological resources in general, including goods and services. Many other national laws and regulations refer to biodiversity and benefit sharing and, thus, there is an explicit effort to understand or apply this principle in a broader manner.

To date, only the ITPGRFA seems to have resulted in an operational, multilateral ABS system. Existing information provided by the Treaty Secretariat indicates that the SMTA has permitted a continued and dynamic flow of resources, in particular from the research centers of the Consultative Group on International Agricultural Research to a wide range of other international and national research agencies. This dynamism can be explained by the fact that the ITPGRFA (its multilateral system) applies to a closed, fixed set of seeds that must circulate unimpaired for research and development and that the SMTA is a standard, non-negotiable instrument that serves this purpose. Parties to the treaty, in exercising their sovereign rights, have decided to place resources within this system and subject them to a mechanism of facilitated access. This contrasts dramatically with the bilateral, case-by-case negotiations that are explicitly promoted by the Nagoya Protocol and all national ABS laws and regulations, and that continue to generate uncertainty, especially at the national level and within the scientific community.

Limitations and challenges: toward a broader view of access and benefit sharing

Benefit sharing, in the context of access to genetic resources, may be leaving out options and possibilities whereby providers and users of genetic resources (a country, institution or community) are, in fact, sharing benefits that derive not from access to and use of genetic resources per se, but from the availability and potential use of biological resources and other products that arise from biodiversity.

A text-based interpretation of the CBD would omit situations where stimulating benefit sharing is not only possible and viable, but highly desirable.

One interesting effect of the CBD has been that, even though ABS is a very focused area, more and more institutions, experts, indigenous peoples and a broad range of actors involved in biodiversity conservation are referring to benefit sharing in much broader, not strictly ABS contexts. Benefit sharing is becoming a concept in itself, not necessarily linked to access or to genetic resources. This view has been summarized in the Addis Ababa Principles and Guidelines for Sustainable Use of Biodiversity (adopted by the Seventh Conference of the Parties), and a number of experts have been trying to put this point of view forward (e.g. see the report on the Informal Expert Workshop on Practical Guidelines for Equitable Sharing of Benefits of Biological Resources in Biotrade Activities (IDDRI 2006)).

A few examples may illustrate this broader point of view, which will be expanded in specific cases in the following sections of this publication. In the context of indigenous peoples and their communities, concepts of distributive justice, reciprocity and equity are well-researched criteria that guide how benefits from management and use of land and resources, barter, labor, seed selection and distribution, participation in decision-making, access to water sources, etc., are shared among community members. Benefit sharing in these situations is triggered by many specific circumstances and, clearly, communities are not basing their actions on CBD obligations, much less ABS frameworks, but rather on traditional and cultural practices (customary law in some cases). Examples of recent experiences in this regard can be found in Suneetha and Pisupati 2009. For a more specific case involving Andean indigenous peoples and benefit sharing at the community level, see Argumedo and Stenner 2008.

Another illustrative example concerns PPB (for a succinct review of the conceptual framework of PPB, see Vernooy 2003). In PPB, farmers, researchers, local consumers and other actors are often involved in a continuing, highly dynamic and complex process of selection and exchange of seeds, interactions between farmers and seed producers (sometimes cooperatives or associations), interactions with research institutions and links with market analysts and, sometimes, with sanitary authorities and government officials. In PPB, benefits occur throughout the process of collaboration and are shared dynamically and at all times among the diverse actors. Ultimately, it is hoped that farmers are the main beneficiaries in terms of new and better-adapted varieties and, maybe, additional income for their seeds or other produce. Here again, the CBD and ABS are not the incentives for PPB; the concept of benefit sharing permeates the PPB process in a broader and more flexible way. Several of the cases described in Part 2 offer more insight into this.

A third area where the concept of benefit sharing is being increasingly invoked is in biotrade projects and initiatives. The United Nations Conference on Trade and Development (UNCTAD) developed and promoted the concept of biotrade in the early 1990s (UNCTAD 2007) and presented the idea during the First Conference of the Parties of the CBD in 1994. Although there is no universal definition of biotrade, the BioTrade Initiative refers to "those activities of collection/production, transformation, and commercialisation of goods and services derived from native biodiversity … under criteria of environmental, social and economic sustainability" (BioTrade Initiative n.d.).

In this case, practitioners and those directly involved in biotrade activities as well as policymakers have recognized that benefits are generated and should be shared equitably, especially with the weakest actors in the research–production–marketing chain: the local or indigenous communities. The elements of equitable benefit sharing in a biotrade project generally include transparency, adequate compensation, non-monetary benefit sharing and empowerment. Biotrade projects also consider traditional knowledge and the implications of its use in research, development, production and commercialization activities.

Finally, benefit sharing could also be considered to be part of the debate surrounding environmental or biodiversity services. In this case, benefits are more extended and their sharing more diffuse. Environmental services analysis is, at present, determining the benefits and costs of maintenance and protection of ecosystems. But in conceptual terms, benefits that are not part of usual ABS discussions are also being shared. The key problem here is the trigger for benefit sharing.

Some economists are suggesting that, as relevant as non-monetary benefits are, potential sharing of monetary benefits should not be overlooked. Writers such as Joe H. Vogel, and some other researchers, have been especially active in reminding policymakers and academics about the informational (intangible) characteristics of genetic resources (including seeds) and the far-reaching economic and con-servation implications this has. Information economics and law offer important lessons that are only just beginning to receive attention.

This specific feature of genetic resources—their informational nature—has been sidelined almost completely in international policy debates and, therefore, an important field for policy and regulatory development has been overlooked. The best-known publication in this area is *Genes for sale: privatization as a con-servation policy* (Vogel 1994). A more recent publication is *Logic should prevail: a new theoretical and operational framework for the international regime on access to genetic resources and the fair and equitable sharing of benefits* (Ruiz *et al.* 2010). Both argue that, if genetic resources are interpreted as shared genetic information and its various expressions (including derivatives resulting from biological metabolism), economic theory calls irremediably for a cartel or monopoly that operates on the basis of a multilateral system where benefit sharing (monetary) is triggered at the point of commercialization of products arising from the use of genetic information and its manipulation (for a detailed discussion, see the Epilogue to this book).

References

Argumedo, A. and Stenner, T. (2008) Association ANDES: conserving indigenous biocultural heritage. International Institute for Environment and Development, London, UK. Gatekeeper series 137a. Available at: www.iied.org/pubs/pdfs/14567IIED. pdf (accessed 1 April 2011).

BioTrade Initiative (n.d.) Selected biotrade definitions and concepts. UNCTAD, Paris, France. Available at: www.unctad.org/biotrade/docs/biotrade-definitions.pdf (accessed 1 April 2011).

Carrisoza, S., Brush, S.B., Wright, B.D. and McGuirre, P.E. (2004) *Accessing biodiversity and sharing the benefits: lessons from implementation of the CBD.* IUCN, Gland and Cambridge.

FAO (Food and Agriculture Organization of the United Nations) (2009) International treaty on plant genetic resources for food and agriculture. FAO, Rome, Italy. Available at: ftp://ftp.fao.org/docrep/fao/011/i0510e/i0510e.pdf (accessed 1 April 2011).

FAO (Food and Agriculture Organization of the United Nations) (2010) *The 2nd state of the world's plant genetic resources for food and agriculture.* FAO, Rome, Italy.

Frison, C., López, F. and Esquinas, J.T. (eds.) *Plant genetic resources and food security: stakeholder perspectives on the International Treaty on Plant Genetic Resources for Food and Agriculture.* Earthscan, London (2011, forthcoming).

IDDRI (Institut du développement durable et des relations internationales) (2006) Informal expert workshop on practical guidelines for equitable sharing of benefits of biological resources in biotrade activities: meeting report. IDDRI, Paris, France. Available at: www.biotrade.org/btfp/BTFP-docs/Reports/BS_workshop_Report.pdf (accessed 1 April 2011).

Ruiz, M., Vogel, J.H. and Zamudio, T. (2010) Logic should prevail: a new theoretical and operational framework for the international regime on access to genetic resources and the fair and equitable sharing of benefits. Sociedad Peruana de Derechos Ambientales, Lima, Peru. Available at: www.spda.org.pe/portal/_data/spda/documentos/2010031611 0250_Serie%2013%20Ingles.pdf (accessed 1 April 2011).

Suneetha, M.S. and Pisupati, B. (2009) Benefit sharing perspectives from enterprising communities. United Nations University, Institute of Advanced Studies, Tokyo, Japan. Available at: www.ias.unu.edu/resource_centre/UNU-UNEP_Learning_from_practitioners.pdf (accessed 1 April 2011).

ten Kate, K. and Laird, S.A. (2002) *The commercial use of biodiversity: access to genetic resources and benefit sharing.* Earthscan, London and Sterling

UN (United Nations) (1992) Convention on biodiversity. United Nations, New York, NY, USA. Available at: www.cbd.int/doc/legal/cbd-en.pdf (accessed 1 April 2011).

UN News Centre (2010) Nations agree on historic UN pact on sharing benefits of world's genetic resources. United Nations News Service, New York, NY, USA. Available at: www.un.org/apps/news/story.asp?NewsID=36618&Cr=biodiversity&Crl (accessed 1 April 2010).

UNCTAD (United Nations Conference on Trade and Development) (2007) UNCTAD biotrade initiative: biotrade principles and criteria. United Nations, New York, NY, USA. Available at: www.unctad.org/en/docs/ditcted20074_en.pdf (accessed 1 April 2011).

Vernooy, R. (2003) *Seeds that give: participatory plant breeding.* International Development Research Centre, Ottawa, Canada. Available at: www.idrc.ca/en/ev-30294-201-1-DO_TOPIC.html (accessed 1 April 2011).

Vogel, J.H. (1994) *Genes for sale: privatization as a conservation policy.* Oxford University Press, Oxford, UK.

3   A brief review of recent access and benefit-sharing initiatives

Ronnie Vernooy and Manuel Ruiz

Work in progress

Over the past decade or so, a number of funding agencies and some development institutions have paid considerable attention to conservation of genetic resources or, more broadly, biodiversity, and these initiatives have covered ABS questions. The result has been a range of projects, some focused on generating research results, others centered on the development of ABS policies, policy measures and legislation. This is a positive development.

There is overall consensus that effective implementation of Article 15 of the CBD requires considerable financial and technical support, and that a long timeframe is required to achieve results. This implies support for timely and appropriate policy and legal research to inform policymaking processes (at the national level in particular, but in large countries, such as Brazil, China and India, at the subnational level as well) and building and strengthening the capacities of institutions and their staff to manage ABS regulations and procedures, monitor their effectiveness and make adjustments as needed. Despite the fact that many countries have now signed on to international agreements and treaties, national processes and mechanisms for their implementation, monitoring and enforcement are often deficient. Policies and laws that recognize more strongly and support more actively the key contributions of rural people to dynamic biodiversity conservation and improvement—and rural innovation more broadly—are still very much a work in progress.

In this chapter, we present a selective overview of initiatives around the world that aim to support fair ABS in practice. We look at some major global projects, followed by a short review of national ABS policy initiatives and of a number of selected policy measures. This review builds on Vernooy *et al.* 2008.

Existing access and benefit-sharing projects

A large, global ABS project, the Genetic Resources Policy Initiative (GRPI), was led by Biodiversity International (formerly the International Plant Genetic Resources Institute) and implemented by national partners in countries, such as Egypt, Nepal, Peru, Uganda, Vietnam and Zambia, and in two subregions,

East Africa and West and Central Africa. The initiative, which operated from approximately 2002 to 2007, was supported by several donor agencies: the German Federal Ministry for Economic Cooperation and Development (BMZ), the Canadian International Development Agency (CIDA), Deutsche Gesellschaft für Technische Zusammenarbeit (GTZ), the International Development Research Centre (IDRC), the Rockefeller Foundation and the Netherlands Ministry of Foreign Affairs. Its main objective was to support transparent, participatory, multidisciplinary ABS decision-making processes and, thereby, build national capacities to implement both CBD principles and specific ABS legislation at the national level. The project resulted in a large number of reports, papers, articles and books. Special attention was paid to highly debated issues, such as farmers' rights and how to implement them effectively. In some cases, specific policies and legislation were developed.

The IDRC-funded project Fair Access and Benefit Sharing of Genetic Resources: National Policy Developments in China, Jordan, Nepal and Peru (2007–11) focused on identifying and supporting practical examples of benefit sharing among indigenous peoples (including farmers) and local communities and other actors involved in crop development. Through multistakeholder dialogue, the project also addressed how national policies and laws could better promote fairness and equity in the distribution of benefits derived from the use of plant genetic resources for food and agriculture (PGRFA). Coordinated by IDRC and Wageningen University, it involved national partners from the public sector and civil society in the participating countries. More details about this project can be found in Part 2 of this book (see the chapters about Peru, Jordan, China and Nepal).

The promotion of a broader look at benefit sharing in farming contexts, specifically in the Hindu Kush-Himalayas region, has also been the objective of a project led by the International Centre for Integrated Mountain Development (ICIMOD) based in Nepal and supported, since 2005, by BMZ and GTZ. This initiative aims to enhance the capacity of governments (in Nepal, India, Bangladesh and Bhutan) at the national, provincial and local levels to: design and implement practical ABS strategies, legislation and guidelines that favor those most in need; improve relations among local, provincial and central levels of governments; and strengthen reciprocal trust and social legitimacy among countries in the region. Involving local, marginalized communities in the design and realization of government policies on ABS and defining their own ABS solutions has received considerable attention.

The Access and Benefit Sharing Research Project coordinated by the Fridtjof Nansen Institute of Norway focuses on implementation of the CBD in Africa, with emphasis on its ABS provisions and principles. It seeks to improve knowledge management and consolidate the conceptual, legal and technical foundations related to ABS. This project is part of the ABS Capacity Development Initiative for Africa (2008–11), a multidonor effort to create awareness of ABS issues and build institutional and individual capacities for ABS policymaking, including such activities as contract negotiations, international negotiations, developing scientific research initiatives and supporting networking in Africa and beyond concerning key policy matters. Funding is provided by GTZ, the Danish Ministry for the Environment, the Norwegian Ministry of Foreign Affairs, the Organisation Internationale de Francophonie, the Institut de l'Énergie de l'Environnement de la Francophonie, and the French Ministry of Foreign Affairs.

This project builds on an earlier initiative that arose from the action plan for capacity-building for ABS adopted by COP 6 in 2002. The Directorate-General for International Cooperation of the Netherlands' Ministry of Foreign Affairs and GTZ funded GTZ's supraregional program, Implementing the Biodiversity Convention. This program took stock of bioprospecting in Africa and assessed ABS capacity development needs, concerning, among other things, the lack of awareness of the potential of ABS at the political level, insufficient national ABS regulations, inadequate implementation of existing regulations, insufficient regional harmonization, and unavailability of inventories and information on the value of genetic resources.

The ABS Project: Law, Equity and Biodiversity, coordinated by the Environmental Law Centre of the International Union for Conservation of Nature and supported by BMZ, was a global initiative to support effective implementation of the ABS principles in the CBD. Its main areas of work included analysis of practical tools and instruments for implementing national ABS frameworks (i.e. certificates of origin or legal provenance), development and improvement of national ABS legal instruments, assessment of intellectual property rights and their relation to the ABS international regime (now the Nagoya Protocol on ABS), among others. The project was initiated in 2003 and ended in 2007. Key research results were compiled in a series of publications produced by international ABS experts (Young 2009).

Another major initiative was the ABS Management Tool: Best Practice Standard and Handbook for Implementing Genetic Resources Access and Benefit Sharing Activities project, led by the International Institute for Sustainable Development. Its main output, the ABS Management Tool, is a best practice standard handbook that provides guidance on ABS practice and tools to help companies, researchers, local and indigenous communities and governments ensure compliance with the Bonn Guidelines and ABS requirements under the CBD (Cabrera 2007). It offers users and providers of genetic resources a structured process for participating in—and making decisions about—ABS negotiations and the implementation of ABS agreements on the use of genetic resources. Volume 1 gives an overview of ABS and the relevance of the ABS management tool for users and providers of genetic resources. It includes best practice standards and advice on key management processes to support its implementation. Volume 2 contains guidance on ABS processes, supporting tools to apply specific aspects of ABS practice and three case studies with additional information on applying the ABS management tool, and lessons learned from field tests of the tool and other ABS processes.

The United Nations University has an ongoing, global ABS program. It focuses on three broad areas of research: evaluation of the effectiveness of existing ABS governance at the international level, identification of capacity development needs to implement ABS policies, and principles and development of research and policy tools. Recent outputs in the form of reports (Suneetha and Pisupati 2009a,b) include an overview of how benefit sharing is actually put in place at the community level in a number of countries and an analysis of policy options to support benefit-sharing mechanisms. The United Nations University (UNU) has also produced in-depth analysis of critical ABS- (and traditional knowledge-) related issues, such as ABS in the context of the Antarctic Treaty, the role of registers in protecting traditional knowledge, the role of customary law in ABS and traditional knowledge-related policy and frameworks, among others.

In the area of genetic resources for food and agriculture in particular, a benefit-sharing fund has been established by the ITPGRFA to support projects in developing countries that ensure on-farm conservation of plant genetic resources and facilitate sharing of benefits from the use of these resources in research, breeding, exchange and in-situ conservation. The financial resources for this fund have been committed by parties to the ITPGRFA (especially Norway, Spain and Switzerland). The first round of grants, awarded in 2009, benefited projects in Peru, Kenya, Morocco, Tanzania, Costa Rica, Nicaragua, Egypt and Uruguay, among others. Parties to the Treaty have also developed a Global Plan of Action (adopted in 1996 and updated regularly since then), which encourages national governments to develop and implement appropriate policy measures at various levels, from community to international (International Technical Conference on Plant Genetic Resources 1996).

Also related to the ITPGRFA, the Global Crop Diversity Trust (established in 2007) focuses on supporting overall conservation of diversity of PGRFA to ensure food security worldwide. This fund receives contributions from a wide range of donors, including companies, countries and financial institutions.

Finally, the Global Environmental Facility provides continued support, guided by CBD COP, to ABS related activities and projects throughout the world.

Development of comprehensive national access and benefit-sharing policies

Among the first countries to attempt to develop and implement holistic, national ABS policies are the Philippines, Costa Rica, South Africa, India and Peru (the work of the latter is highlighted in Chapter 5). Currently, more than 50 countries are developing ABS or ABS-like policies (Table 3.1).

The GRPI is providing technical support to a number of countries around the world engaged in this process, notably Egypt, Zambia, Nepal, Vietnam and Peru. The GRPI also operates at a subregional level in West, Central and East Africa. It uses a multistakeholder, multidisciplinary and multisectoral approach (the "3M approach") to bring underappreciated social actors (such as farmers) into the policy process, to encourage scientists, farmers, lawyers and policymakers to join forces, and to build bridges across sectors, between agriculture and environment, for example. It is hoped that this process of capacity building and collaborative work (including research) will lead to more effective genetic resource policies

Table 3.1 Selected national policies and laws, including drafts

Policy or law	Country	Year	Objective
Executive Order 247 on Access to Biological and Genetic Resources	The Philippines	1996	Regulates access to biological and genetic resources and traditional knowledge of indigenous peoples
Decision 391 on a Common Regime on Access to Genetic Resources	Andean Community (Bolivia, Colombia, Ecuador, Peru)	1996	Regulates access to genetic resources, derivatives and traditional knowledge of indigenous peoples
Law 7788, Biodiversity Law	Costa Rica	1998	Regulates access to biological and genetic resources and protects traditional knowledge of indigenous peoples
Model Law on Rights of Local Communities, Farmers, Breeders and Access	African Union (53 African states)	2000	Regulates access to biological and genetic resources and traditional knowledge of local communities and small farmers, as well as access to and use of seeds by breeders
Framework Agreement on Access to Genetic and Biological Resources (draft)	Association of South Eastern Asian Nations (ASEAN)	2000	Regulates access to biological and genetic resources and traditional knowledge of indigenous peoples
Provisional Measure on Access to Genetic, Biological Resources and Traditional Knowledge	Brazil	2001	Regulates access to biological and genetic resources and traditional knowledge of indigenous peoples
Biological Diversity Act	India	2002	Addresses conservation and sustainable use of biodiversity and contains specific references to genetic resources
Bonn Guidelines and Access to Genetic Resources and Benefit Sharing Arising from their Use	Global (non-binding, soft law)	2002	Regulates access to biological and genetic resources, including through user measures and intellectual property rights
Biodiversity Act	Bhutan	2003	Addresses conservation and sustainable use of biodiversity and contains specific references to genetic resources
Central American Agreement on Access to Genetic and Biological Resources and Traditional Knowledge (draft)	Central American Commission	2003	Regulates access to biological and genetic resources and traditional knowledge of indigenous peoples

Policy or law	Country	Year	Objective
International Treaty on Plant Genetic Resources for Food and Agriculture	Global (binding)	2003	Regulates conservation and sustainable use of PGRFA, including access and benefit sharing through the multilateral system
Access to Genetic Resources and Community Knowledge and Community Rights Proclamation No. 482/2006	Ethiopia	2006	Regulates access to biological and genetic resources and traditional knowledge of indigenous peoples
Executive Decree 257 on Access to Genetic Resources and Traditional Knowledge	Panama	2006	Regulates access to biological and genetic resources and traditional knowledge of indigenous peoples
Supreme Decree 003-2009-MINAM, Regulation on Access to Genetic Resources	Peru	2009	Regulates access to genetic resources and derivatives and traditional knowledge of indigenous peoples
Nagoya Protocol on Access to Genetic Resources and the Fair and Equitable Sharing of Benefit Arising from the Use	Global (binding)	2010	Regulates access to genetic resources and derivatives and the use of traditional knowledge of indigenous peoples

and, hence, to the sustainable use of genetic resources. According to recent reports (GRPI 2007, 2008), the five years of support provided by GRPI has led to improved policymaking processes and outcomes in the targeted countries, assessed in terms of increased awareness, more coherent conceptual frameworks for policy development, a number of actual policy measures and the inclusion of genetic resources policy issues in higher education curricula. Detailed country reports are not yet available, making it difficult to assess field-level impacts.

A limited number of studies have tried to assess the development and implementation of ABS policies, for example, in terms of stakeholder participation (Swiderska 2001 (four countries)), and the effectiveness of ABS (e.g. Richerzhagen and Holm-Mueller 2005, Miller 2006 (Costa Rica)). Costa Rica has been relatively successful in developing an ABS policy through a balanced strategy dealing with key impact factors, such as property and intellectual property rights, rules and practices of enforcement and bargaining among various stakeholders. Bioprospecting (coordinated by a national intermediary organization, the Instituto Nacional de Biodiversidad), has been at the core of the policy, but, more recently, policy measures have been added in other areas, such as ecotourism and payment for environmental services (Richerzhagen and Holm-Mueller 2005: 452). The study does not specifically address the impact on PGRFA or agricultural biodiversity.

Case study: the Philippines

The Philippines was among the first countries to develop a comprehensive ABS policy. In May 1995, then-president Ramos signed Executive Order 247, "Prescribing guidelines and establishing a regulatory framework for the prospecting of biological and genetic resources, their by-products and derivatives, for scientific and commercial purposes, and for other purposes." The order covered all forms of bioprospecting and, hence, of all kinds of biological resources, including PGRFA, whether found on public, private or ancestral lands. As in Costa Rica, a coordinating intermediary agency, the Inter-Agency Committee for Biological and Genetic Resources, was established to implement the order. The process leading to the order was initiated and led by a group of concerned scientists, who wished to ensure that the exploitation of Filipino biodiversity had direct benefits to the country. A national consultation process involving academics, NGOs and government agencies was carried out to draft, discuss and agree on the final text. According to Swiderska *et al.* (2001: 7–8), who analyzed this process, the consultation process was generally praised for being fairly broad and comprehensive, but it also had some limitations: it was rather *ad hoc* and limited to the capital, key government officials were not consulted and concerns expressed by some organizations were excluded.

A case study of the order focusing on impact concludes that "the regulation provided a sophisticated system for benefit sharing, covering a wide variety of benefits and beneficiaries over a broad time span. In a few cases there were some examples of benefit sharing, but no actual or potential benefits were achieved with regard to plant genetic resources for food and agriculture" (Andersen 2007: 28–9). The author speculates that the order may not have been widely known and, therefore, not followed or that the regulation was too demanding in terms of plant breeding and, therefore, ignored. She concludes that, after all, the order "was much ado about nothing." Although the strategy used to design and adopt an ABS policy was successful, in the end it did not lead to effective and sustained implementation, largely because one crucial stakeholder group (the bioprospectors) was not taken into consideration. This important lesson is also highlighted by Swiderska *et al.* (2001).

Policy measures

Farm and community level in-situ conservation

In situ refers to on-site, the area where plants or animals have developed their distinctive properties, i.e. in the wild or in farmers' fields. While ex-situ or off-site conservation efforts (in a gene bank or botanical garden) are stable or declining, in-situ conservation initiatives of various types have sprung up around the globe, often led by national or international NGOs in cooperation with local communities and frequently supported by scientists. These in-situ efforts are responding to the widely accepted fact that agricultural intensification is contributing to a widespread decline in farm and community-level biodiversity (Brookfield *et al.* 2003). Formal, government-led, in-situ conservation policies have made much slower progress, although in some countries, such as Nepal, a deliberate attempt has been made in recent years to support in-situ efforts more strongly.

The variety of in-situ conservation initiatives has considerably increased our understanding of their merits and challenges. This has helped answer key questions, such as: What do women and men farmers know about the properties and uses of agricultural genetic resources (including conservation and improvement) and how can this knowledge be respected, strengthened and used appropriately and fairly for the benefit of local communities and wider society? What are viable management practices, fair cost, benefit-sharing mechanisms and useful incentives to strengthen in-situ conservation and improve agricultural genetic resources under conditions of agro-ecological and socioeconomic change? (see, for example, Brush 2000, Brookfield *et al.* 2003, Vernooy 2003, Jarvis *et al.* 2007).

A number of major research projects have been instrumental in this regard, e.g.: the global Community Biodiversity Development and Conservation project; the IPGRI-led in-situ conservation of agrobiodiversity on farms project; the Local Initiatives for Biodiversity, Research and Development (LI-BIRD)-led in-situ conservation project in Nepal (see Chapter 11); the global People, Land Management and Environmental Change project; the Seeds of Survival Program in Ethiopia; the efforts of the Centro Internacional de la Papa—Users' Perspectives With Agricultural Research and Development (CIP-UPWARD) concerning sweet potatoes in the Philippines; the efforts of the Centre for Biodiversity and Indigenous Knowledge in Yunnan province, China; the Biodiversity Use and Conservation in Asia and Participatory Enhancement of Diversity of Genetic Resources (PEDIGREA) projects in South and Southeast Asia; several efforts carried out in India including the work done by the Honey-Bee Network and the Green Foundation; the Center for Chinese Agricultural Policy's work in Guangxi province; the work led by the Instituto Nacional de Ciencias Agricolas in Cuba (many of these examples are documented in CIP-UPWARD 2003, Volumes 1–3). The IPGRI-led project, in particular, merits attention as it was one of the first initiatives to include in-situ conservation as an integral part of a national plant genetic resources program. As such, it aimed to contribute directly to policy formulation and implementation.

Providing economic and regulatory support to local organizations is a key part of in-situ conservation. Local organizations comprising farmers and other interested parties (including government officials) can mobilize local resources (land, water, labor, seeds, funds) in individual farm and community units to increase access to new genetic resources and build the exchange and utilization networks required to maintain dynamic conservation processes. Mburu and Wale (2006) discuss cases of on-farm conservation of traditional cereals and pulses in Ethiopia and indigenous vegetables in Kenya. They point out that the conservation of genetic resources by local organizations can be classified according to certain factors that influence interactions among stakeholders and the devolvement of decision-making authority, in particular access to markets, the presence of collective action or self-organizational capacity and the establishment of relevant conservation policies. They highlight the importance of policies that support marketing of traditional crops, such as investment in infrastructure, awareness-building campaigns and the removal of adverse subsidies. However, they caution that policies based on market access may have a negative effect on biodiversity, as others have observed, as farmers might be tempted to concentrate on crops with high market value.

Research into on-farm conservation in northeast Zimbabwe points in a similar direction, cautioning that too much emphasis on marketing may be detrimental to household food security (Cromwell and Van Oosterhout 2000). In this study, the authors suggest that governments could develop policies to support investment in processing equipment for various crops, which would lessen the burden on women and free up time to cultivate a wider variety of crops. In a study on seeds, seed politics and gender in southern India, Pionetti (2005: xv–xvi) draws attention to the gendered nature of policies as well. She concludes that the growth of the commercial seed sector has had a profound impact on local seed exchange systems. In areas where commercial crops have almost completely displaced food crops (her case study focuses on the Deccan Plateau), the practice of saving seeds is disappearing, with subsequent loss of local knowledge of agrobiodiversity, traditional breeding, seed selection, seed production and storage. By losing their prerogative over seeds, women have lost their main means of ensuring mixed cropping in their fields, with adverse consequences for the land and for plant diversity. Our review revealed very few significant gender-informed policies, on paper, let alone in practice.

Protecting farmers and traditional/indigenous resource rights

Designing and implementing appropriate and effective measures to protect traditional, indigenous or local rights (e.g. as embodied in the concept of farmers' rights) to PGFRA has been a major challenge during the period under review. This may be due to the fact that these rights, and the practices they aim to protect, are the very basis of the sustainable use and management of genetic resources. There continues to be a hotly contested debate about these rights, especially on the international scene, although often without the involvement of the rightholders concerned, i.e. representatives of indigenous, local farmer communities or organizations (Kuyek 2002; Vernooy 2003; Hardon 2004: 41). Only a relatively small number of national governments have tried to design and implement meaningful policy measures that are clearly farmer-centered (instead of focused on plant breeders); India is one of them (see case study below; see also Brush 2007: 1509–1510) and Nepal is another (Sharma 2005).

In Africa, the Organization for African Unity (OAU) developed the *African Model Legislation for the Protection of the Rights of Local Communities, Farmers and Breeders, and for the Regulation of Access to Biological Resources* (1998) to guide national governments to craft specific national legislation (Kuyek 2002). The OAU African model legislation has been praised for its clear vision and strong commitment to protecting the rights of indigenous and farmer communities (Mushita and Thompson 2002; Zerbe 2002; Brush 2007). Governments that are now trying to implement national policies based on the model legislation have set the stage for a more equitable distribution of benefits associated with biodiversity and biotechnology. However, actual implementation remains a challenge (Zerbe 2002: 317). Indigenous and farmer communities have an important role to play, but often remain excluded from the policy and implementation process (Kuyek 2002: 18).

> Indigenous plants, not the global market, provide Southern Africa with food security; in marginal environments where varied nutritious crops provide insurance against the failure of one crop and in periods of drought … . In Southern Africa, biodiversity is a resource of the poor, the majority, and there is organized resistance to its being privatized. Southern Africa civil societies and governments are not simply rejecting the "universality" of the WTO TRIPS, but are working to pass legislation that offers political and legal alternatives. The legislative draft, calling for local and national control, could be a model for other countries to transform the incongruities between TRIPS and the CBD into complementarities.
>
> (Mushita and Thompson 2002: 80)

In contrast to this pan-African initiative, a number of interesting, national and local-level policy "experiments" (formal and less formal) have been underway across the globe that could perhaps pave the way for national-level design and implementation. We review several of these in the case studies.

Case study: India's *Protection of Plant Varieties and Farmers' Rights Act*

The *Protection of Plant Varieties and Farmers' Rights Act* (PPVFR), approved in 2001, relates both to the protection of farmers' varieties of seed via the *sui generis* option outlined in the Agreement on Trade Related Aspects of Intellectual Property Rights (TRIPS) and to other international agreements and treaties, such as the ITPGRFA. The objective of the PPVFR as stated in its preamble is to establish "an effective system for the protection of plant varieties, the rights of farmers and plant breeders, [and] to encourage the development of new varieties of plants." According to Ghose (2003; see also Gene Campaign 2007: 122–34; Dutfield 2008: 46), the PPVFR attempts to strike a balance, satisfying both the concerns of farmers over their ability to acquire, save and sell seed, and the concerns of breeders who desire adequate protection for their research and resultant technologies. According to Gene Campaign (2007: 137), the PPVFR is the first legislation in the world to grant farmers formal rights without jeopardizing their self-reliance. With regard to the sharing of financial resources that result from the successful commercialization of local knowledge or the transfer of local varieties to state or private parties for breeding, the PPVFR introduced a National Gene Fund. Its purpose is to collect funds for the original holders of the genetic resource. Unfortunately, we have not been able to determine what the fund has achieved so far (see Dutta 2005).

    Although the PPVFR became law in April 2002, India has applied for accession to the International Union for the Protection of New Varieties (UPOV). This move seems contradictory, especially with respect to farmers' rights, a central component of the PPVFR. According to Ghose (2003: 22), "Given that the only version of UPOV that potential members can be party to is the 1991 version, and that this version has made 'plant back rights' an exception, it is unlikely that the two can coexist with respect to farmers' rights." Dutfield (2008: 47) goes even further, stating that it has become difficult for any developing country to design and implement its own system of plant variety protection, given the politics involved, nationally and internationally.

Developing the agricultural value chain: promoting agro-tourism

Adding commercial value to local agricultural biodiversity is emerging as a means to maintain dynamic systems of genetic resources. Some countries are experimenting with policies to attract tourists to experience agricultural biodiversity (in the landscape and on farms) and to get them to pay for the experience in one way or another, e.g. through the purchase of services (entrance fees, tourist guidance) or local products, foods and specialties. Especially in countries of the South, agro-tourism has the potential to increase awareness of agricultural biodiversity and the farming systems that maintain it, increase communication between farmers and visitors, and strengthen the links in value chains or create new links and chains.

However, much has yet to be learned. A recent review of experience to date (GTZ 2007a) concluded that, "In order to market the local attraction successfully, the involvement of other bodies may be necessary—marketing agencies for the development of tourism products and advertising strategies; tourism associations for the distribution of information, to serve as a contact point and to make arrangements with guests; and local and regional planners to ensure that the infrastructure is adapted to tourist needs."

Assigning protected designations

Another way to add value to agricultural biodiversity is by assigning so-called protected designations, and the European Union (EU) has been at the forefront in formalizing this type of intervention. In 2006, to promote regional and product-specific diversification and provide better protection for distinctive cultural features, the EU introduced a series of designations, such as: "protected geographical indication," to highlight the geographic origin of a product (although the processing could take place elsewhere); "protected designation of origin," for products that are processed at their place of origin; and "traditional specialty," awarded to products and food made from raw materials or using a traditional process. Traditional means that special, local knowledge must have been used and transmitted over at least one generation (GTZ 2007b). Of interest is the fact that not only European producers and manufacturers may register products, but that non-Europeans are also entitled to do so.

The GTZ review provides two examples from the South (GTZ 2007a). A case in Mexico concerns the designation of a seal of origin to mezcal, a liquor made of the agave plant. This has led to positive results in terms of both livelihoods and biodiversity conservation, as a large number of agave varieties can be used to produce mezcal. However, in the case of a Vietnamese rice variety called Tam Xoan, the results have been mixed: successful in terms of livelihoods, but detrimental to biodiversity as the focus on Tam Xoan has led farmers to neglect other rice varieties. This tendency has also been observed by other researchers. For example, Kruijssen *et al.* (2007: 24), who looked at five underutilized plant species in four countries (Thailand, India, Vietnam and Syria) point out that links between local efforts, especially collective action, and enhanced on-farm management of biodiversity may be indirect, taking place through a variety of networks that differ across contexts.

Recently, some researchers have proposed the use of designations to protect crop varieties to ensure collective innovation, keeping access to relevant germplasm open to farmers and arranging for fair benefit sharing. Salazar *et al.* (2007: 1525–6) argue that geographic indications are adequate as a model or guide, because they allow all producers who make products in a designated place to use them. Geographic indications are also protected in accordance with national laws as well as a number of international treaties, including WTO-TRIPS. The authors cite the red rice varieties of Bohol in the Philippines as an example. These varieties could be designated as protected and marketed as local farmer varieties by naming them Bohol Red Paddy. This would be a clear example of the recognition of collective efforts, which, after all, are the basis for maintaining and improving genetic resources.

Across the world, there are examples of crop varieties that have been designated or are in the process of being designated as organically produced, through one or more certification schemes. This is another alternative to add value to genetic resources.

As the examples in this chapter indicate, in policy and legal terms, ABS is still a relatively new concept. Thus, trial and error implementation efforts and continuing social and technical research are helping to clarify and consolidate ABS as a key principle in the search for global fairness and equity in the use of biodiversity and its components.

References

Andersen, R. (2007) Governing agrobiodiversity: plant genetics and developing countries. Paper prepared for the 48th ISA Convention, Chicago, February 28–March 3, 2007. Fridtjof Nansen Institute, Lysaker, Norway.

Brookfield, H., Parsons, H. and Brookfield, M. (eds.) (2003) *Agrobiodiversity: learning from farmers around the world.* United Nations University Press, Tokyo, Japan.

Brush, S.B. (2000) The issues of in situ conservation of crop genetic resources. In: S.B. Brush (ed.) *Genes in the field: on-farm conservation of crop diversity.* Lewis Publishers, Boca Raton, FL, USA, International Development Research Centre, Ottawa, Canada, and International Plant Genetic Resources Institute, Rome, Italy. pp. 3–28.

—— (2007) Farmers' rights and protection of traditional agricultural knowledge. *World Development* 35 (9): 1499–514.

Cabrera, J. (2007) ABS-management tool: best-practice standard and handbook for implementing genetic resources access and benefit sharing activities. International Institute for Sustainable Development, Winnipeg, Canada. Available at: www.iisd.org/pdf/2007/abs_mt.pdf (accessed 14 March 2011).

CIP-UPWARD (Centro Internacional de la Papa—Users' Perspectives With Agricultural Research and Development) (2003) *Conservation and sustainable use of agricultural biodiversity: a sourcebook.* CIP-UPWARD, Los Baños, Laguna, Philippines.

Cromwell, E. and Van Oosterhout, S. (2000) On-farm conservation of crop diversity: policy and institutional lessons from Zimbabwe. In: S.B. Brush (ed.) *Genes in the field: on-farm conservation of crop diversity.* Lewis Publishers, Boca Raton, FL, USA; International Development Research Centre, Ottawa, Canada; International Plant Genetic Resources Institute, Rome, Italy. pp. 217–38.

Dutfield, G. (2008) Turning plant varieties into intellectual property: the UPOV Convention. In: G. Tansey and T. Rajotte (eds.) *The future control of food: a guide to international negotiations and rules on intellectual property, biodiversity and food security.* Earthscan, London and Sterling, UK. pp. 27–47.

Dutta, M. (2005) Priorities for farmers' rights in India. *Farmers' Rights* 1(1): 4.

Gene Campaign (2007) Protection of indigenous knowledge of biodiversity: a report. Gene Campaign, New Delhi, India.

Ghose, J.R. (2003) *The right to save seed.* International Development Research Centre, Ottawa, Canada. Unpublished paper.

GRPI (Genetic Resources Policy Initiative) (2007) Strengthening capacity to analyze national options (annual technical report). Bioversity International, Rome, Italy.

—— (2008) Strengthening capacity to analyze national options (annual technical report). Bioversity International, Rome, Italy.

GTZ (Deutsche Gesellschaft für Technische Zusammenarbeit) (2007a) Creating value from products with protected designations (People, food and biodiversity issue papers). GTZ, Eschborn, Germany.

—— (2007b) Maintaining and promoting agricultural diversity through tourism (People, food and biodiversity issue papers). GTZ, Eschborn, Germany.

Hardon, J. (2004) *Plant patents beyond control: biotechnology, farmer seed systems and intellectual property rights.* Agromisa, Wageningen, the Netherlands. AgroSpecial 2.

International Technical Conference on Plant Genetic Resources (1996) Global plan of action for the conservation and sustainable utilization of plant genetic resources for food and agriculture. Food and Agriculture Organization of the United Nations, Rome, Italy. Available at: http://typo3.fao.org/fileadmin/templates/agphome/documents/PGR/GPA/gpaeng.pdf (accessed 14 March 2011).

Jarvis, D.I., Padoch, C. and Cooper, H.D. (eds.) (2007) *Managing biodiversity in agricultural ecosystems.* Columbia University Press, New York, USA, and Bioversity International, Rome, Italy.

Kruijssen, F., Keizer, M. and Giuliani, A. (2007) Collective action for small-scale producers of agricultural biodiversity products. CAPRi working paper no. 71. International Food Policy Research Institute, Washington, DC, USA. Available at: www.ifpri.org/sites/default/files/publications/CAPRIWP71.pdf (accessed 14 March 2011).

Kuyek, D. (2002) *Intellectual property rights in African agriculture: implications for small farmers.* GRAIN, Barcelona, Spain.

Mburu, J. and Wale, E. (2006) Local organizations involved in the conservation of crop genetic resources: conditions for their emergence and success in Ethiopia and Kenya. *Genetic Resources and Crop Evolution* 53: 613–29.

Miller, M.J. (2006) Biodiversity policy making in Costa Rica: pursuing indigenous and peasant rights. *Journal of Environment and Development* 15(4): 359–381.

Mushita, A. and Thompson, C.B. (2002) Patenting biodiversity? Rejecting WTO/TRIPS in Southern Africa. *Global Environmental Politics* 2(1): 65–82.

—— (2007) *Biopiracy of biodiversity: global exchange as enclosure*. Africa World Press, Trenton.

Pionetti, C. (2005) *Sowing autonomy: gender and seed politics in semi-arid India*. International Institute for Environment and Development, London, UK.

Richerzhagen, C. and Holm-Mueller, K. (2005) The effectiveness of access and benefit sharing in Costa Rica: implications for national and international regimes. *Ecological Economics* 53: 445–60.

Salazar, R., Louwaars, N.P. and Visser, B. (2007) Protecting farmers' new varieties: new approaches to rights on collective innovations in plant genetic resources. *World Development* 35(9): 1515–28.

Sharma, P. (2005) Options to protect farmers' rights in Nepal. *Farmers' Rights* 1(1): 5.

Suneetha, M.S. and Pisupati, B. (2009a) Benefit sharing perspectives from enterprising communities. United Nations University, Institute of Advanced Studies, Tokyo, Japan. Available at: www.ias.unu.edu/resource_centre/UNU-UNEP_Learning_from_ practitioners.pdf (accessed 14 March 2011).

—— (2009b) Benefit sharing in ABS: options and elaborations. United Nations University, Institute of Advanced Studies, Tokyo, Japan. Available at: www.ias.unu.edu/sub_page. aspx?catID=111&ddlID=884 (accessed 14 March 2011).

Swiderska, K. (2001) Stakeholder participation in policy on access to genetic resources, traditional knowledge and benefit sharing: case studies and recommendations. International Institute for Environment and Development, London, UK. Available at: http://pubs.iied.org/9096IIED.html (accessed 14 March 2011).

Swiderska, K., Daño, E. and Dubois, O. (2001) Developing the Philippines' Executive Order No. 247 on access to genetic resources. International Institute for Environment and Development, London, UK. Available at: http://pubs.iied.org/9061IIED.html (accessed 14 March 2011).

Vernooy, R. (2003) Supporting agricultural biodiversity conservation: key questions. In: CIP-UPWARD *Conservation and sustainable use of agricultural biodiversity: a sourcebook*. International Potato Center-Users' Perspectives with Agricultural Research and Development, Los Baños, Laguna, pp. 33–8.

Vernooy, R. with Li Jingsong and Zhang Li (2008) The impact of national, regional and global agricultural policies and agreements on conservation and use of Plant Genetic Resources for Food and Agriculture. A contribution to the 2nd State of the World's Plant Genetic Resources for Food and Agriculture. Food and Agriculture Organization of the United Nations, Rome. Available at: www.fao.org/docrep/013/i1500e/i1500e19.pdf (accessed 14 March 2011).

Young, T.R. (2009) Covering ABS: addressing the need for sectoral, geographical, legal and international integration in the ABS regime: papers and studies of the ABS project (IUCN environmental policy and law paper 67/5). International Union for Conservation of Nature and Natural Resources, Gland, Switzerland. Available at: http://data.iucn.org/ dbtw-wpd/edocs/EPLP-067-5.pdf (accessed 14 March 2011).

Zerbe, N. (2002) Contested ownership: TRIPs, CBD, and implications for Southern African biodiversity. *Perspectives on Global Development and Technology* 1 (3–4): 294–321.

Part 2

Practical experiences from around the world

4 Introducing the case studies—access and benefit sharing in practice

Ronnie Vernooy and Manuel Ruiz

Although ABS questions have been debated at the international level for a decade, the concept remains relatively new in the world of rural development research—in the broad sense, including both natural and social science-oriented studies. There is relatively little guidance for planning and implementing feasible ABS mechanisms, which remain a matter of experimentation. Indeed, this is still an area for pioneers!

Around the world, expertise has been building in a number of initiatives, but the sharing of results, failures, insights and lessons has been minimal. Thus, there is scope for more exchange to build a field of knowledge and experience to better inform current and future practices, policies and laws, and to prevent the same mistakes. Examining links and opportunities for shared learning with a wider set of partners, including "non biodiversity" partners, such as those working on common pool resources and more practitioner-oriented organizations, may also be useful, as may be examining mechanisms to promote non-monetary benefit-sharing approaches related to developing livelihood resilience. For most farmers, genetic resources are a means to an end, not an end in themselves

Case studies are one way to deepen the field of knowledge and experience. In the following chapters, we present a number of them from Latin America and the Caribbean, the Middle East and Asia. These cases represent long-term, field-based efforts to address key issues and ABS questions in practice. All are characterized by their direct engagement with policy- and law-making processes at the national level, bringing field experiences, insights and lessons to the attention of key decision-makers. Some case study team members participate directly in national committees or platforms on behalf of their organization or constituency (in Peru, Cuba, Jordan, Nepal and China). Some team members have been and continue to be active in regional or international fora dealing with ABS (in Peru and Nepal).

The research initiatives described in the case studies are concerned with agricultural biodiversity and related traditional knowledge and practices. Agricultural biodiversity, although under threat around the world for a number of reasons (FAO 2010), remains at the heart of farmers' livelihoods, especially those of many indigenous and ethnic minority peoples and communities in the Andean region, Himalayas, southwest China and elsewhere. In all cases, except Peru, teams have been at the forefront in introducing, testing and adapting participatory plant breeding (PPB), most for more than a decade now. The Peru case represents an example of a champion in the field of biopiracy study, awareness raising and policy influence, complementing the other cases by using a different but important lens.

The teams all have a common goal: improving farmers' and indigenous peoples' livelihoods by ensuring food security, better food quality, improved well-being, support for local cultural and collective identities, the dynamic use and maintenance of biodiversity and collective capacity for innovation.

Core development research issues

The case studies address, to various degrees and in various forms, four interrelated core development and research issues.

Improving the quality of genetic resources

- Defining a clear rationale and approach with a balance between farmer-focused outcomes and scientist-focused outcomes
- Documenting and analyzing findings
- Ensuring cooperation between natural and social sciences
- Using quantitative and qualitative methods
- Paying attention to cost–benefit analysis
- Recognizing local practices and knowledge: doing research in farmers' fields
- Recognizing genotype versus environment interactions and the great variability in local contexts: doing experiments in farmers' fields
- Identifying and acknowledging various forms of participation; agreeing on appropriate incentives and compensation
- Developing a formal or informal code of conduct
- Taking shared authorship seriously
- Over time, expanding the focus from single crops to multiple cropping systems.

Exploring viable and fair seed production, distribution and marketing to improve farmers' individual and collective livelihoods

- Using a differentiated approach to self-pollinated and cross-pollinated crops
- Developing new methods to assess contributions to the development of varieties arising from PPB (recognition of the collective nature of local innovation processes)
- Paying attention to social recognition, e.g. appropriate and fair naming of varieties, awards for individual and collective efforts
- Respecting cultural values concerning exchange, diffusion and commercialization
- Improving access, for example, by establishing local gene banks and linking in-situ conservation to ex-situ conservation
- Developing viable economic and benefit-sharing arrangements

- Exploring appropriate non-economic incentives and benefit sharing
- Recognizing and learning how to deal with the politics of recognition and ABS.

Scaling out and scaling up: building bridges between local practices and supra-local policies, laws and agencies that have a rural development mandate

- Strengthening cooperation between local/indigenous peoples and researchers, including farmer–plant breeders cooperation; involving farmers early; designing flexible experiments
- Recognizing farmers fields as "research stations" in national and agency agendas
- Supporting farmers' organizational efforts at local and supra-local levels
- Supporting farmer-to-farmer capacity development
- Acknowledging the roles of women and supporting them
- Using new extension approaches
- Building links between research and teaching; supporting curriculum innovation
- Developing individual and organizational facilitation capacities.

Creating an enabling policy and legal environment: the political economy matters

- Critiquing existing (or non-existing) policies and laws governing recognition and ABS, including free-trade agreements and other political–economic pacts
- Paying attention to the economics of ABS
- Paying attention to the science and technology driving research in genetic resources
- Dealing with policy vacuums; bringing local realities into the policy domain
- Addressing concerns about all types of genetic resources and related knowledge (e.g. landraces, medicinal plants)
- Addressing issues of intellectual property rights
- Dealing, through alternative mechanisms, with the bottleneck caused by inadequate implementation of policies and enforcement of laws
- Exploring a focus on innovation policies.

Key research and development questions

The ABS theme covers a number of related, complex issues that require holistic and dynamic research approaches and, because of this complexity, a sufficiently long time horizon. Some of the key research and development questions that the research teams are dealing with are:

- How do people at the local level perceive and assess ABS questions, especially in light of national and international guidelines, model laws and other new forms of defining and regulating ABS regarding biodiversity resources?

- How can local and indigenous knowledge and practices be acknowledged, recognized and valued? How can the principles of prior informed consent and mutually agreed terms (e.g. in the case of model agreements), including the settlement of possible disputes and remedies and arbitration, be respected?
- How can the roles and responsibilities and the forms of participation of right-holders and stakeholders be defined (e.g. through formal or informal codes of conduct)?
- What are the means to ensure respect for, conserve and strengthen indigenous/local knowledge, customary practices and innovations? How should questions of intellectual property rights and ownership of genetic resources and related knowledge be dealt with? What are appropriate incentives and how can they be used?
- How can feasible ABS mechanisms, both formal and informal, be designed, implemented and monitored? How can conflicts between local-level ABS priorities and national/international interests be avoided? How can existing conflicts be resolved? How can conditions be created to reduce future conflicts?

A synopsis of the case studies

Peru

Since the CBD came into force, Peru has been very active in developing frameworks for ABS and protecting traditional knowledge, because of its strong concern over the misuse, illegal use and misappropriation (biopiracy) of national genetic resources and indigenous peoples' traditional knowledge. This case study offers an overview of the implementation of and some of the overarching lessons learned from the project Action Oriented Research and Activities to Support Implementation of Access and Benefit Sharing Policies and Laws in Peru: Confronting Misuse and Misappropriation of Seeds, Genetic Resources and Traditional Knowledge (part of a larger, IDRC-supported project on ABS of genetic resources).

Syria

Syria was a pioneer in the Middle East in PPB, a local action research process in which farmers and breeders are engaged in a collaborative learning process. This approach enables farmers to benefit from their contributions to the global genetic pool, for example, by adding value to their crops, improving their livelihoods and increasing their income. The PPB work in Syria served later as a learning ground for similar efforts in other countries in the region, e.g. Algeria, Morocco, Egypt, Jordan and Yemen. PPB is one of the most common types of benefit sharing related to farmers' rights as outlined in the ITPGRFA.

Jordan

Jordan's government has been developing a supportive institutional environment for the country's agricultural sector. This case study describes the relevant agricultural policies, laws and international agreements, and how they were enacted, through the lens of the country's efforts to introduce and institutionalize PPB. The PPB efforts build on those of Syria and other countries. ABS issues are still very new to the country, but are gaining recognition. This case is also part of the larger, IDRC-supported ABS project.

Honduras

In April 2006, Honduras entered into a free-trade agreement with the United States: the Dominican Republic–Central American Free Trade Agreement (DR-CAFTA). This agreement will have a profound effect on smallholder agriculture, on which a large percentage of the population in the poorest countries of the region depends. This case study examines the impact of the agreement on ABS of plant genetic resources for small Honduran farmers and their seed systems, through the lens of farmers' rights. The study concludes that farmers' rights are unlikely to be defensible, notwithstanding the ITPGRFA. Under DR-CAFTA, the seeds of farmers and indigenous people and related knowledge are considered patentable commodities, and obtaining prior informed consent, disclosing origin and ensuring sharing of benefits before patent application are considered unnecessary.

China

In marginalized areas of China, such as the southwest part of the country, farmers' seed systems continue to play a major role in the seed supply system and maintaining the diversity that is essential to sustain the livelihoods of all farmers and the country at large. This case study presents the experience of a decade of efforts to link community-based action research on the conservation of agricultural biodiversity and crop improvement (maize in particular) with relevant policy- and law-making processes at the national level by engaging key decision-makers in the rural development policy arena at local, provincial and national levels. Preliminary results of a series of novel policy experiments at the local level are presented, and the implications for national policies and laws are discussed. Although results have been positive, there is still insufficient attention to farmers' contributions to maintenance and improvement of genetic resources and their rights in general. This is the third case of the IDRC-supported ABS project.

Cuba

Cuban agriculture is struggling to survive under difficult conditions, as is much of the country's economy. Farmers across the island, together with a number of young agricultural researchers, are rediscovering that necessity is the mother of invention. They are trying to breathe new life into agriculture by reviving its heart: the seed systems that are the basis for production and reproduction. In the process, new forms of participation and cooperation have emerged and, through these, new

ABS arrangements are evolving. This is happening not because of a predesigned plan, but as an expression of guiding principles that are informing the remaking of the seed systems, principles based on a more flexible, open and dynamic view of how social change can be brought about. This case study provides an overview of the revitalization process.

Nepal

Until the beginning of the 1990s, ABS was an unknown concept in Nepal. There was no recognition that an ABS regime could form the basis for the protection of the rights of local, indigenous and farming communities over genetic resources and related traditional knowledge. It was only after Nepal became a party to the CBD, in February 1993, that the government and several NGOs began to discuss the importance of integrating ABS issues into national policies. This case study describes innovative research and development efforts to give concrete meaning to the concept of ABS and to create a policy and legal environment promoting the diversity, both biological and sociocultural, on which Nepal depends. This project, which is also part of the IDRC-supported ABS project, is called Promoting Innovative Mechanisms for Implementing Farmers' Rights through Fair Access to Genetic Resources and Benefit Sharing Regime in Nepal.

Reference

FAO (Food and Agriculture Organization of the United Nations) (2010) *The 2nd state of the world's plant genetic resources for food and agriculture*. FAO, Rome, Italy.

5   Peru

Seeking benefit sharing through a defensive approach—the experience of the National Commission for the Prevention of Biopiracy

Manuel Ruiz

Located on the west coast of South America, Peru covers an area of 1,285,215 km^2 broadly divided longitudinally into an arid coastal region and the very diverse Andes and Amazon regions. It contains more than 80 recognized life zones, making it one of ten mega-diverse countries of the world. It is a center of origin of agriculture and of some of the most important food crops in the world, including the potato, as well as many varieties of maize, hot peppers, Andean roots and tomatoes. It is also home to over 70 groups of indigenous peoples, mainly Quechua and Aymara in the Andes, but very diverse and many unique groups in the Amazon. Culture and diversity are closely related: new crops, medicinal plants, domesticated animals, natural dyes and a wide array of other natural products have been discovered, developed and put to use by Andean and Amazon communities. These communities continue to develop, adapting to new conditions such as climate change, and broader society has also benefited (Brack Egg 2003).

This case study focuses on how misuse, illegal use or misappropriation of genetic resources and traditional knowledge (biopiracy) affect benefit sharing and the cultural and social interests of the country, especially given new ABS principles recognized in the CBD. Furthermore, through careful research, biopiracy cases have been documented, analyzed and brought to the attention of national and international audiences (Ruiz 2005).

Unlike the other case studies, the experience in Peru is not based on participatory plant breeding or in-situ activities, but on a broader issue that is a priority for public policy and legislation: proper compliance with national ABS principles and regulations. This issue cuts across all the other case studies presented in this publication and may serve to inform and assist efforts in Nepal, Jordan, China (all part of the IDRC's ABS project) and other countries that are looking for balance and equity in how their genetic resources and traditional knowledge are accessed and used.

The Peruvian policy and legal context

The Andean Community was the first regionally integrated bloc to adopt a comprehensive policy and legal framework for access to genetic resources and protection of traditional knowledge, as a pioneering step in implementing the equity and fairness principles of the CBD. Decision 391 of the Andean Community on a Common Regime on Access to Genetic Resources (1996) regulates who may have access to the region's genetic resources and under what conditions. It also establishes general obligations for the recognition and protection of traditional knowledge. The Andean Community, which was formed in 1969, currently includes Bolivia, Ecuador, Colombia and Peru. Decisions made by the community are binding on all member countries (Caillaux *et al.* 1999).

Decision 391 was adopted because countries in the region realized that their biodiversity wealth was being used by industry and biotechnology firms, with very limited compensation to the countries or their people. Misappropriation and biopiracy were an important part of the discussions that led to Decision 391. At the time, the media had documented a series of high-profile cases, in which inventions based on biodiversity from the region had been patented, much to the surprise and opposition of the Andean countries and their indigenous peoples. In the 1990s, patents on ayahuasca (*Banisteriopsis caapi*) and later quinoa (*Chenopodium quinoa*) highlighted the vulnerability of the region, in terms of their sovereign rights and interests over native biodiversity. At the same time, this situation provoked further tensions and intensified the debate and discussions between the industrialized and technologically advanced North and the biodiversity-endowed South.

Since that time, Peru has been very active in developing its own policy and legal framework related to genetic resources and protection of traditional knowledge. Two important regulations are Law 27811, the *Law for the Protection of the Collective Knowledge of Indigenous People* (2002), and the recently enacted Supreme Decree 003-2009-MINAM (2009) which regulates Decision 391. Both seek to implement CBD principles and establish specific norms and provisions for the protection of traditional knowledge and access to genetic resources, respectively (Venero 2005: 219).

Although efforts to implement these regulations have been made by the National Institution for the Defense of Competition and Intellectual Property (INDECOPI) and the Ministry of the Environment respectively, there is still much to do in terms of strengthening institutional capacities to enforce them. Implementation requires efforts that not only focus on restrictions and limitations, but also create appropriate incentives and, in general, ensure that access as provided under the CBD is facilitated. Research should not be impeded. Furthermore, there is a need to ensure that traditional knowledge is used in a manner that serves the research process, but that prior informed consent is obtained and the specific interests of indigenous peoples are considered.

Other factors complicate implementation of the regulations at the national level. Over the past decade or so, new technological and scientific advances have been changing how research and development related to biodiversity components takes place. Bioinformatics, genomics, proteomics, synthetic biology and genetic engineering have revolutionized how encoded natural information in genes and other molecular structures can be read, manipulated and transformed into useful products in almost all sectors of human activities (for a detailed discussion, see

Pastor and Ruiz 2009). As a result, decades-old legal frameworks and even current templates do not appropriately apply to these new research paradigms.

An important development that is also a result of initial activities related to Decision 391 has been the enactment of specific defensive provisions, using the intellectual property rights system to prevent biopiracy by establishing appropriate conditions for processing patent applications. Decision 391 includes complementary provisions that stipulate that member states will not recognize intellectual property rights over inventions based on activities that have not complied with regional access legislation. Furthermore, patents are not granted in such cases (Correa 2005).

Decision 486, the *Common Regime on Industrial Property* (2001), goes even further. It makes granting of patents conditional on the presentation of clear and detailed contracts and instruments showing evidence of compliance with legislation governing access to resources and traditional knowledge. In brief, these defensive protection measures ensure that patents are not issued without evidence (in the case of biodiversity-related inventions) that resources and traditional knowledge used in the invention were obtained legally and in compliance with national access and traditional knowledge protection laws.

Not only have specific laws in Peru, Brazil, Costa Rica, India, Switzerland and Norway included defensive protection principles and obligations, but calls have consistently been made in international forums such as WIPO and the World Trade Organization (WTO) to modify patent disclosure requirements and add other requirements to safeguard countries' biodiversity and traditional knowledge interests. Calls for disclosure are now high on the political agenda of many international institutions and organizations. Disclosure of origin and legal provenance of genetic resources and traditional knowledge is seen as an appropriate way to create positive synergy between the intellectual property system and the CBD ABS principles and, thus, contributes to a coherent international policy and legal architecture (Henninger 2010).

Creation of the National Commission for the Prevention of Biopiracy

Since 1996, INDECOPI has been leading the country in efforts to enact a national law for the protection of traditional knowledge. The agency was also very much involved in the Andean Community's Decision 391 development process. In 2002, INDECOPI took notice of a case where a patent had been granted to Pure World Botanicals for an invention based on a Peruvian plant, maca (*Lepidium meyenii*) (US patent 6428824). The media had reported extensively on this case. Thus, its general features were widely accessible on the Internet, although details of the actual patent claims and the background of the invention were not given.

INDECOPI convened a working group to look into this case in detail and determine whether biopiracy was indeed an issue. The working group agreed to review the patent documents to determine whether maca had been accessed legally, whether traditional knowledge was in some way involved in the invention and whether the invention complied with the universal criteria for patentability (novelty, inventiveness and industrial application). Legal and technical assistance for this work in the United States was provided by the organization Public Interest Intellectual Property Advisors.

During 2002–03, the working group realized that this was not the only patent on maca; not only were there many other patents related to the same plant, but there were also hundreds of patents (in various technological fields) related to a wide range of plants of Peruvian origin. As a result, the Commission to the Peruvian Congress proposed transforming the working group into a longer-term body that would continue to address cases of biopiracy.

In 2004, Congress enacted Law 28416 which created the National Commission for the Prevention of Biopiracy to look into biopiracy cases since 2001. Law 28216 defines "biopiracy" as:

> non authorized nor compensated access to and use of biological resources or traditional knowledge of indigenous people by third parties, without the appropriate authorization and in contravention to the Convention on Biological Diversity principles, and existing laws. This appropriation may take place through physical control, intellectual property rights over products which include illegally obtained elements or often through intellectual property claims.

Thus, in Peru, biopiracy is legally defined and includes non-compensation (an element of benefit sharing) as a cause for initiating administrative or judicial action.

The Commission is made up of representatives from public and private organizations, such as INDECOPI, the National Institute for Natural Products, the Peruvian Society for Environmental Law, the National Institute for Indigenous Peoples, the National Institute for Agricultural Innovation, the ministries of the Environment, Agriculture and Foreign Relations and others. Its main functions are to prevent biopiracy, identify patents and patent applications that might involve biopiracy, analyze these patents and take action against them, prepare national positions to present at international fora where biopiracy-related issues may be discussed (i.e. CBD, WIPO's IGC and the WTO) and support communities in their efforts to combat and address biopiracy.

Over the past few years, the Commission has been successful in three areas. First, it has provided an ideal arena for national institutions with different interests and views to come together and discuss a national position on biopiracy; as a result, the Commission has achieved coherence and a united position. Second, the Commission has put biopiracy on the national agenda, and the country leads international CBD, WIPO and WTO interventions in terms of technical analysis and proposals for preventing and addressing biopiracy.

For example, Peru presented to the IGC an analysis of potential cases of biopiracy related to a set of ten Peruvian plants over which patent applications and rights had been processed and granted (IGC 2005). This document generated many responses from industry and others, e.g. a submission from the Biotechnology Industry Organization and the International Federation of Pharmaceutical Manufacturers on regional, national and community policies, measures and experiences regarding intellectual property and genetic resources (IGC 2010a).

Awareness raising has been a central element in the Commission's work. Workshops and seminars on biopiracy prevention have been held in at least seven regions of Peru, convening local authorities, indigenous peoples and community representatives and members. A documentary on biopiracy (illustrating the maca case and one involving sacha inchi) has also been produced recently by the Commission and the Peruvian Society for Environmental Law.

The Commission's website (www.biopirateria.gob.pe), which is continually updated, contains useful and detailed information. A practical manual, *How not to be a biopirate in Peru*, has also been produced with support from the Commission. The experience of the Commission has also been recognized in other countries; for example, the Ecuador Institute for Intellectual Property received a capacity-building course to allow it to learn from the experience of Peru's Commission and adapt it to its own situation. In addition, collaboration is being sought between the Commission and India's Traditional Knowledge Digital Library.

Third, at least seven cases of biopiracy have been successfully fought through a series of actions, including intervention by the Ministry of Foreign Affairs, negotiations with foreign companies and communications with patent offices in Japan, France and the United States.

Examples of biopiracy and their implication for benefit sharing

Over the past few years, the National Commission has identified and taken action against a series of patents, which, under Law 28216 and on considerable legal and technical analysis, have been deemed to be cases of biopiracy. The following are some examples.

- In the case of Japanese patent 2007031371, "Ameliorant for Sleep Disturbance," the claim concerns an alcoholic extract of maca that prevents sleep disturbances. The National Commission sent the Japanese Patent Office information related to the use of maca as a regenerator. The patent was subsequently abandoned in 2009 by Suntory Ltd and the University of Hiroshima.
- Japanese patent application 2005306754 (2005) "Testosterone Increasing Composition" by Towa Corporation includes a claim on compounds extracted from maca—benzyl glucosinolate and benzyl isothiocyanate—that increase the level of testosterone in the blood. The Commission sent the Japanese Patent Office relevant information regarding traditional uses of maca in Peru, and the application was subsequently rejected and the case closed in 2008.

- In 2009, the Ministry of Foreign Affairs informed the National Commission that Greentech Company (France) was withdrawing its patent related to sacha inchi (*Plukenetia volubilis*), a native Peruvian plant. This came after the Commission had provided technical information to the French patent authorities arguing against the novelty and inventiveness of the patent. This case also received attention from the French Collectif de Biopiraterie and the French parliament.
- Also in 2009, the National Commission received a letter originating from Cognis IP Management GmbH Company indicating its decision to withdraw its European patent application 05802707.9 (2006) on an extract of sacha inchi and its cosmetic use because of evidence submitted by the Commission. In other words, the company acknowledged not having added anything substantially new to the use of the resource.

In all these cases, the novelty and inventiveness of the proposed patents was arguable. For many years, indigenous and local communities have been aware of the properties and uses of these substances, even though they have not expressed them in Western scientific terms nor necessarily documented them in books or formal research papers.

In late 2009, the National Commission contacted the Molecular Biophysics and Biochemistry Department at Yale University requesting information regarding its permits for research and collection of endophytic microorganisms in the region of Puerto Maldonado (in the Amazon). Yale indicated that this research was part of ongoing collaboration with the Vargas Herbarium in Cusco and denied any wrongdoing in terms of biopiracy. The case is being closely monitored by the Commission, as this is the first case in which biopiracy is being addressed in the context of possible illegal access to resources rather than wrongful patents.

These are just a few examples of the types of cases addressed by the National Commission. The Commission's main goal is not to act as a policing force, although at this early stage it is necessary to do so. Its ultimate objective is to demonstrate, through careful technical analysis and examples, that the international patent system is flawed in the sense that it grants rights when it should not. This stems from the very nature of the system—the grant first, receive challenges later attitude prevalent in many patent offices. Solving the problem will require better patent searches, collaboration between patent offices and biodiversity management offices, and international disclosure requirements as a precondition for processing a patent application. Opportunities for benefit sharing and research should be promoted through sound policies and laws (Suneetha and Balakrishna 2009).

However, lack of respect for the social, cultural and economic interests of countries in the South in their biodiversity, traditional knowledge and genetic resources undermines the possibility of developing appropriate international legal frameworks to protect their biodiversity and its components. For Peru, bringing this to the attention of the world remains a top priority.

From a reactive to a proactive approach to seeking equitable benefit sharing

The role of the National Commission for the Prevention of Biopiracy has been twofold. First, as the discussion above demonstrates, it has been responsible for preventing biopiracy with regard to Peruvian biodiversity and the traditional knowledge of its indigenous peoples. Second, it has contributed to identifying best practices for realizing benefit sharing.

Over the past six or more years, as it has examined specific cases of biopiracy, the commission has accumulated considerable experience regarding how best to create and enhance an institutional environment that monitors how Peru's biodiversity is being used, while creating appropriate incentives to stimulate investment in research and development activities. Over the past few years, the Commission has made special efforts to raise awareness among a wide range of stakeholders about the rules governing access to biodiversity components and the use of traditional knowledge of indigenous people. Workshops, field trips, international participation in meetings and interactions with national and foreign companies have helped the Commission to act less like police and to address how best to facilitate research and development in relation to national biodiversity.

However, the country is still far from establishing a fully operational framework for protecting access to genetic resources and traditional knowledge. Local examples of effective ABS mechanisms could play an important role in refining such an operational framework. One such example is the Potato Park in the mountains of Cusco (see box). This story shows how local, collective action can inspire national (and even international) policy- and law-making.

The Potato Park in Cusco

The Potato Park is an agrobiodiversity zone located in the Pisac area of the Andes. It is an ideal spot for native potatoes, and a wide range of Andean tubers and roots, including their wild relatives, are grown there. It is a *sui generis* protected area that covers a "collective bio-cultural heritage site." In 2001, the Andes Association, an NGO based in Cusco, joined six farm communities in the Pisac area (Paru Paru, Chawaytire, Cuyo Grande, Cuyo Chico, Sacaca, Pampallacta) to form the Association of Communities of the Potato Park, covering an area of 10,000 ha (IIED 2007).

The Potato Park was created to protect the biocultural heritage of the campesino communities. This includes protecting their traditional knowledge, preventing biopiracy, developing registers of local biodiversity and traditional knowledge, promoting ecotourism, repatriating lost crops (with support from the International Potato Center) and promoting the sale of local products, including soaps, shampoos and medicinal plants. The aim is to provide communities with a development option based on their own needs and interests, using market forces to satisfy these interests. Critically important are the cultural, spiritual and ancestral customs as elements guiding livelihoods and activities within the park (IGC 2010b).

Two important areas of work have been the prevention of biopiracy and calling for compliance with appropriate ABS rules and principles. The park has developed a local ABS protocol to guide collecting of material for crops or medicinal purposes within its area. For example, in 2004, the Potato Park communities publicly denounced a biopiracy case regarding nuña, a native crop, when a patent was awarded to a US food processing company (Appropriate Engineering and Manufacturing).

The Potato Park has also stimulated policy development at the regional level. The Regional Government of Cusco enacted an ordinance (Regional Ordinance 010-2007) calling for action against biopiracy and establishing a moratorium on the introduction of genetically modified crops into Cusco agricultural systems.

The Potato Park is an example of a positive and proactive approach to implementing the ABS principles of the CBD and national legislation, but at a regional and local level, more grounded on specific local needs and expectations.

It has created positive synergies among various levels of government and a wide range of institutions. For example, the regional government created a Regional Biopiracy Prevention Group, which is interacting with the National Commission, and an agrobiodiversity technical working group, which is developing its own agrobiodiversity action plan for Cusco. The six communities of the Potato Park have also been empowered and are now more involved in developing a broader range of regional policies that affect their livelihoods, seed exchange, biosafety and agrobiodiversity.

Source: Potato Park website: www.parquedelapapa.org

Efforts are being made to recognize and support the creation of "agrobiodiversity zones"—areas where there is particularly high genetic diversity and a strong Andean and Amazonian cultural presence. The Potato Park is an example. Supreme Decree 068-2001-PCM regulates the recognition and creation of these areas. Agrobiodiversity zones are meant to support and enhance local livelihoods through seed conservation and protection of local culture and practices. At the same time, they seek to create incentives for communities to undertake sustainable activities that enable them to link and interact with markets. Strengthening local seed exchange systems, organic agriculture and agro-ecotourism are just three ways in which new opportunities are created for communities to consolidate their livelihoods. Local biodiversity registers are used as a tool to prevent biopiracy while offering communities a degree of control over their resources and helping them revalue their traditional rights and heritage.

Conclusions

Biopiracy is a widespread phenomenon, often difficult to detect, but sometimes blatant and obvious. It is certainly not universally recognized as illegal, although it has been defined as such in Peruvian legislation. Biopiracy is a concept with strong political connotations, and this has been extremely important in convincing countries of the need to conserve and regulate how components of their biodiversity are used. Biopiracy is occurring in many places and it is expected to spread. Small communities around the world have few, if any, resources for denouncing, let alone combating, biopiracy.

It is because of biopiracy that the benefits gained from accessing and using genetic resources and traditional knowledge have not been equitably shared between commercial or industrial users and those who provided the resources and knowledge. Biopiracy is the reason behind efforts to secure fairer terms regarding the flow of genetic resources (and related traditional knowledge), especially from the South to the North. At the same time, it is clear that monitoring and policing these flows can be costly. Thus, while these actions are critically important, the National Commission advocates a broader strategy, based on well-established and functional policies and laws governing access to resources and protecting traditional knowledge. The Peruvian experience to date has much to offer other countries.

Biopiracy has a detrimental effect on social, cultural and even the economic interests of countries and especially communities. There is a sense of loss and frustration that cannot be easily overcome, especially when this practice takes place most often in foreign jurisdictions. Investment in monitoring how resources are used and the cost of legal and administrative action also make combating biopiracy very complicated.

In Peru, expectations are high that the recently approved Nagoya Protocol concerning access to and benefits from genetic resources offers an opportunity to develop specific rules that can be put into practice and that can prevent biopiracy, but also facilitate access to resources, stimulate research and ensure that benefits are shared effectively. Furthermore, the IGC process offers an important opportunity to ensure that traditional knowledge is protected at the international level.

References

Brack Egg, A. (2003) *Perú: diez mil años de domesticación*. Proyecto FANPE, Lima, Peru.

Caillaux, J., Ruiz, M. and Tobin, B. (1999) *El Régimen Andino de Acceso a los Recursos Genéticos: lecciones y experiencias*. World Resources Institute and Peruvian Society for Environmental Law, Lima, Peru.

Correa, C. (2005) Alcances jurídicos de las exigencias de divulgación de origen en el sistema de patentes y derechos de obtentor. Iniciativa para la prevención de la biopiratería. Documentos de Investigación. Peruvian Society for Environmental Law and International Development Research Centre, Lima, Peru. Year I, no. 2, August.

Henninger, T. (2010) Disclosure requirements in patent law and related measures: a comparative overview of existing national and regional legislation on IP and biodiversity, in *Triggering the synergies between intellectual property rights and biodiversity*. Deutsche Gesellschaft für Technische Zusammenarbeit, Eschborn, Germany, pp. 293–326. Available at: www.cisdl.org/biodiv/gtz2010_IPR_Biodiv_Reader.pdf (accessed 1 April 2011).

IGC (Intergovernmental Committee on Intellectual Property and Genetic Resources, Traditional Knowledge and Folklore) (2005) Patent system and the fight against biopiracy—the Peruvian experience (WIPO/GRTKF/8/12). World Intellectual Property Organization, Geneva, Switzerland. Available at: www.wipo.int/edocs/mdocs/tk/en/wipo_grtkf_ic_8/wipo_grtkf_ic_8_12.pdf (accessed 1 April 2011).

—— (2010a). Joint BIO and IFPMA submission on regional, national and community policies, measures and experiences regarding intellectual property and genetic resources (WIPO/GRTKF/IC/16/INF/21/annex 1). World Intellectual Property Organization, Geneva, Switzerland. Available at: www.wipo.int/edocs/mdocs/tk/en/wipo_grtkf_ic_16/wipo_grtkf_ic_16_inf_21.doc (accessed 1 April 2011).

—— (2010b) Policies, measures and experiences regarding intellectual property and genetic resources: annex 1 (WIPO/GRTKF/IC/16/INF/13). World Intellectual Property Organization, Geneva, Switzerland. Available at: www.wipo.int/edocs/mdocs/tk/en/wipo_grtkf_ic_16/wipo_grtkf_ic_16_inf_13.pdf (accessed 1 April 2011).

IIED (International Institute for Environment and Development) (2007) *Traditional resource rights and indigenous people in the Andes*. IIED, London, UK. Available at: http://pubs.iied.org/pdfs/14504IIED.pdf (accessed 1 April 2011).

Pastor, S. and Ruiz, M. (2009) The development of an international regime on access to genetic resources and fair and equitable benefit sharing in a context of new technological developments. Initiative for the prevention of biopiracy. Peruvian Society for Environmental Law, Lima, Peru. Year 4, no. 10, April.

Ruiz, M. (2005) ¿Cómo prevenir y enfrentar la biopiratería? Una aproximación desde Latinoamérica. Iniciativa para la Prevención de la Biopiratería. Documentos de Investigación. Peruvian Society for Environmental Law and International Development Research Centre, Lima, Peru. Year I, no. 1, January.

Ruiz, M., Vogel, J.H. and Zamudio, T. (2010) Logic should prevail: a new theoretical and operational framework for the international regime on access to genetic resources and the fair and equitable sharing of benefits. Sociedad Peruana de Derechos Ambientales, Lima, Peru. Available at: www.spda.org.pe/portal/_data/spda/documentos/20100316110250_Serie%2013%20Ingles.pdf (accessed 1 April 2011).

Suneetha, M.S. and Balakrishna, P. (2009) *Benefit Sharing in ABS: Options and Elaborations*. Institute of Advanced Studies, United Nations University, Yokohama, Japan. Available at: www.ias.unu.edu/resource_centre/UNU_ABS_Report_Final_lowres.pdf (accessed 1 April 2011).

Venero, B. (2005) Mitos y verdades sobre la biopiratería y la propiedad intelectual, in B. Kresalja (ed.), *Anuario Andino de derechos intelectuales*. Palestra Ediciones, Lima, Peru.

6 Syria

Participatory barley breeding—farmers' input becomes everyone's gain

Salvatore Ceccarelli, Alessandra Galié,
Yasmin Mustafa and Stefania Grando

Kherbet El Dieb, north of Aleppo, is one of 24 Syrian villages involved in a participatory plant breeding (PPB) initiative started by the International Center for Agricultural Research in Dry Areas (ICARDA). Yields there have increased since the farmers have begun using varieties developed through the PPB program. PPB is one of the most common types of benefit sharing related to farmers' rights as the concept is outlined in the International Treaty on Plant Genetic Resources for Food and Agriculture. Combining farmers' knowledge with that of professional breeders, this approach enables the farmers to benefit from their contribution to the global genetic pool by adding value to their crops, improving their livelihoods and increasing their incomes. However, as the name indicates, the main principle of PPB is participation, and this is a signature characteristic of the barley breeding initiative in Syria.

Fawaz Al-Abboud Al-Hassoun, a farmer in Kherbet El Dieb who took part in the project, is very happy with the participatory approach and the resulting varieties. The productivity of the new varieties is high because of their increased resistance to drought and cold and, thus, they have been adopted by many of the farmers in the village.

This case study describes how PPB evolved in Syria and how benefits have been generated through local action research in which farmers and breeders are engaged in a collaborative learning process. The PPB work in Syria also served as a learning ground for PPB in other countries in the region (e.g. Algeria, Egypt, Eritrea, Ethiopia, Iran, Jordan, Morocco and Yemen). An example of this spreading of PPB in Jordan is described in the next chapter.

Participatory research and plant breeding

In recent years, there has been increasing interest in participatory research in general and in PPB in particular, as scientists have become more aware of how users' participation in technology development may increase the probability of success. The interest in PPB stems partly from the view that the impact of agricultural research, including plant breeding, has been below expectations, particularly in developing countries, in marginal environments and among poor farmers. In fact, according to the World Food Programme (WFP 2011), there are 925 million malnourished people in the world today. The limited impact of most agricultural research in marginal areas is, to some degree, due to the fact that the research agenda is usually determined by the scientists and not discussed with farmers. Agricultural research is also typically organized according to disciplines or commodities and seldom adopts an integrated approach that would more closely resemble the situation at the farm level. There is a large gap between the number of technologies generated by the agricultural sciences and the relatively small number adopted and used by farmers, particularly smallholders.

In relation to plant breeding, most scientists would agree that programs have not been very successful in marginal environments or among poor farmers. It takes a long time (about 15 years) to release a new variety, and few of these are adopted by farmers, many of whom grow varieties other than the officially released ones. Even when new varieties are acceptable to farmers, the seed may not be available or it may be too expensive. Great loss of biodiversity is also associated with conventional plant breeding, and reversing this trend is important both to improve the livelihoods of farmers and to maintain plant genetic diversity.

Defined as a type of research in which users are involved in the design—not merely the final testing—of a new technology, participatory research is now seen by many as a way to address these problems. PPB in particular, a plant breeding system that involves scientists, farmers and other partners (such as extension staff, seed producers, traders, consumers and NGOs), in the development of a new variety, is expected to produce varieties that are: targeted at the right farmers; relevant to their real needs, concerns and preferences; and appropriate in terms of producing varieties that will be adopted.

The science behind participatory and conventional plant breeding is the same. The major difference is that conventional plant breeding is a process where priorities, objectives and methods are all decided by scientists, whereas PPB gives equal weight to the opinions of farmers (and other stakeholders). It is also important to distinguish between PPB and farmers' breeding practices, defined as the various complex activities farmers engage in on their own, with no participation by scientists.

Since the beginning of agriculture and until the rediscovery of Mendel's laws and the start of scientific plant breeding, farmers have planted, harvested, stored and exchanged seeds, modified their crops, moved crops around, and, as a result, have been able to feed themselves and the rest of society. Implicit in the way farmers bred their crops was selection for specific adaptations, both to their environment (climate and soil) and their uses. This led to a large number of landraces of all the main crops. During this process, farmers have accumulated an immense wealth of knowledge.

However, at the beginning of the last century, plant breeding was gradually removed from farmers' hands, with the result that what had been done by many, many people in many diverse places was being done by fewer and fewer people in relatively few places. Selection for specific adaptations was replaced by selection for wide adaptation to allow seed companies to multiply and sell a few varieties of seed over large geographic areas.

The wealth of knowledge accumulated by farmers over millennia was not taken into consideration. The difference between traditional knowledge and modern science is probably one of the reasons for this. The former is based on repeated observations over time, whereas the latter is based on repeated observations over space (replications). While traditional knowledge is usually shared informally, modern science is almost always communicated in a written and highly formal manner. Because it is difficult for scientists to elicit traditional knowledge using the forms of communication of modern science, farmers' knowledge has often been ignored or misinterpreted in conventional plant breeding, with the result that the technologies produced did not reflect the local needs and priorities of farmers. In contrast, PPB starts with recognition of farmers' knowledge and expertise, and is concerned with building on it and strengthening it.

The first phases of Syria's participatory plant breeding project

ICARDA, which is one of the 15 international agricultural research centers that make up the Consultative Group on International Agricultural Research (CGIAR), has been involved in PPB in Syria since 1995. PPB is well suited to ICARDA's objective of improving the livelihoods of resource-poor people in dry areas by enhancing food security, alleviating poverty through research and partnerships to achieve sustainable increases in agricultural productivity and income, and ensuring efficient and equitable use and conservation of natural resources. The General Commission for Scientific and Agricultural Research (GCSAR), the formal national research institution for breeding in Syria, was also involved in the PPB initiative from the beginning.

The main goal has been to develop a way to move from top-down centralized breeding programs to bottom-up participatory, decentralized programs. An additional goal was to provide a model that could be used in other countries and for other crops. This is a continuing effort, with 24 villages all across Syria now involved. The widespread nature of the program has been possible partly because of collaboration among GCSAR staff at research stations in the provinces and extension staff who have easy access to farmers in the various villages. Most of these villages are located in marginal areas, frequently affected by droughts and resulting crop losses. The breeding of varieties that are adapted to this climate is, therefore, an important aspect of the project.

Farmers have been involved in PPB from the beginning. At first, this meant consultations not only about the overall objectives but also about organization of the trials (number of varieties, plot size, seeding rate, scoring methods, etc.). Together, participants decided that developing new and better barley varieties in farmers' fields with farmers' participation would be the main priority.

In the beginning, the main objectives were to build relationships (the team), understand farmers' preferences, measure the efficiency of farmers' selection methods, develop a scoring system and enhance farmers' skills. Exploratory work included the selection of farmers and test sites, and the establishment of a common experiment in nine villages and two of ICARDA's research stations. The nine villages represented a range of climatic conditions from wet to dry as well as a range of farmer literacy levels, farm sizes (about 5–160 ha), farm types in terms of the extent of crop and livestock production, levels of income (on-farm and off-farm) and differences in the importance of barley in the farming system. None of the villages had adopted modern varieties even though farmers knew about them and, in some cases, had tried planting them.

Kherbet El Dieb is one of the driest villages selected to participate in the PPB project, with an average annual rainfall of 174 mm. As sheep are the main agricultural product, barley as the main livestock feed plays a critical role in the livelihood of the village. Barley is used solely as animal feed (mainly for sheep) throughout Syria. However, although it might be the only crop choice in dry areas, it is also grown as a rainfed crop in more complex farming systems together with wheat, lentils, chickpeas and summer crops. Farmers with their own herds of sheep will use the barley they grow as feed and sell the surplus, while farmers without herds will sell their entire barley harvest (both grain and straw).

The two participating research stations, Tel Hadya and Breda, are located in two distinct production environments. Tel Hadya, with an average annual precipitation of 338 mm, has a typical high-input, favorable environment for barley and a wide choice of crops. At Breda, on the other hand, with average annual precipitation of 268 mm, the environment is low-input, high-risk; barley is the most common rainfed crop and there is a limited choice of other crops and cropping systems.

The initial barley experiment took place over three cropping seasons (1996–97, 1997–98 and 1998–99) and included 200 new barley types that represented a wide range of characteristics, such as plant height, flowering and maturity date, leaf colour, row type (two vs. six rows), seed colour (white, black, grey), stem diameter and associated lodging resistance and straw palatability. Because barley is used exclusively as animal feed in Syria, straw palatability is a valuable trait for the farmers but is usually neglected by breeders. In addition, eight farmer cultivars from eight of the nine host farmers were also included.

The 208 varieties could be sorted into various categories. They came from either modern germplasm (100) or landraces (108); they were fixed lines (100) or segregating populations (108); they had two rows (158) or six rows (50) and they had white seeds (161), black seeds (28) or mixed seed colours (19).

Both before and after planting, agronomic management of the trials was left to the host farmers. The trials were conducted under rainfed conditions in the farmers' fields as well as at the research stations to ensure that they were grown under typical farm conditions. (At the time, the government did not allow irrigation of barley.)

Each of the participating farmers was given a field book in which to record daily rainfall and observations. Most farmers preferred a numeric scale as a scoring method, while some preferred qualitative scoring, classifying plots as "bad," "medium," "good," "very good" and "excellent." Eventually, they adopted a mix of quantitative scores for some traits and qualitative descriptors for others. The farmers used these scores during final seed selection to assign an overall score. Farmers did not usually need assistance with scoring, but where there was a high degree of illiteracy, they were assisted in recording their scores by other farmers or by the scientists.

Selection processes

Various selection processes were used. Centralized non-participatory selection was carried out by a scientist, in this case GCSAR's barley breeder, at the research station, while centralized participatory selection was conducted by farmers at the research station. The decentralized process was also either non-participatory (carried out by the breeder in the farmers' fields) or participatory, with selection done by farmers in their fields.

The first selection took place in May 1997. The work was done independently by the various participants, who did not know what the others had selected. The varieties were identified based on who selected them and the location from which they were selected:

- selected by farmers in their field
- selected by farmers at Tel Hadya research station
- selected by farmers at Breda research station
- selected by the breeder in each of the farmers' fields
- selected by the breeder at Tel Hadya research station
- selected by the breeder at Breda research station.

The first four groups were specific to the nine farmers' fields, although a number of samples were commonly selected in more than one farmer's field. Using the selected samples and taking care to avoid duplication, a specific trial was prepared for each of the nine farmers' fields. The samples in the two last groups were common to all trials.

In the 1997–98 cropping season, the farmers chose local landraces and improved varieties to use as controls. Abdu Sheiko, a farmer from the area near Al Bab (a large village 60 km northeast of Aleppo) had introduced a forage legume crop into rotation. The trial crop was, therefore, planted twice, once after barley and once after the legume. All ten trial crops were also planted at the two research stations, using the same layout as in the farmer fields. The total number of samples tested in 1998 was 1,348, of which 196 were genetically different as a result of the large diversity reflected in the selection criteria used in 1997. The process of evaluation and selection conducted in 1997 was repeated in 1998 on the lines selected the first year, and again in 1999 on the lines selected in 1998.

Experience during the first three years of the trials indicated that farmers are able to handle large numbers of samples (a frequently debated issue among PPB practitioners), make a number of observations during the cropping season and develop their own scoring methods. It was also observed that farmers select for specific adaptive traits and, in some cases, selection is driven mainly by environmental adaptation. Diversity of farmers' selections was greater in their own fields than at the research stations and greater than those of breeders at both locations. The selection criteria used by the farmers were nearly the same as those used by the breeders. In addition, in their own fields, farmers were slightly more efficient than the breeders in identifying the highest-yielding varieties. The breeders were more efficient than the farmers in selection at the research station located in a high-rainfall area, but less efficient at the research station located in a low-rainfall area. These findings constitute a strong argument for farmer participation.

Benefits

The first phase of the barley PPB project in Syria led to increased awareness among the farmers of the nature of plant breeding and what it can offer. This was evident from the number and quality of questions raised by the farmers during the entire process. Requests to extend PPB to other crops also showed how interested the farmers were in this approach. The fact that farmers were at least as efficient as breeders when it came to selection was an important finding that allowed the approach to be extended to other countries (Algeria, Egypt, Eritrea, Ethiopia, Iran, Jordan, Morocco, Tunisia and Yemen) often after visits by scientists from these countries to Syria during the first project phase.

The demonstrated ability of farmers to handle a large number of populations discredited the belief that they are simple-minded people, incapable of dealing with more than 20–30 varieties at a time. This was essential if the project was to move from the linear process used in the first phase to a cyclic process and a truly participatory program. The results from the three-year experiment indicated that there was much to gain, and nothing to lose, from implementing a decentralized PPB program; thus, a second phase was initiated. This meant ensuring that the farmers knew that the project would not be short term, but ongoing and evolving. The farmers were agreeable, and the project could continue.

The second phase of the project

An important feature of the second phase was that the role of the research stations changed; they were now used only for seed multiplication, making crosses and preparing the initial material. The number of villages taking part in the project increased from 9 to 11 in 2003 and to 24 in 2005. The number of farmers directly involved also increased as a result of strong support from the Syrian Ministry of Agriculture and Agrarian Reform following a workshop organized in Hama at the request of the minister of agriculture. In addition, seed production was initiated in some villages. Details of the experiments, such as the number of lines to be tested, plot size, type of germplasm, selection criteria and issues related to seed production, were discussed in meetings with farmers in each of the participating villages. This led to the development of a more refined PPB model, which ICARDA would subsequently use in other countries.

It is worth mentioning that there are no fixed models for PPB. For a particular crop, even within the same country, different models may be required depending on the genetic structure of the varieties and how farmers are used to handling on-farm genetic diversity, among other factors. In the model used by ICARDA for a number of self-pollinated crops (barley, bread wheat, durum wheat, lentils and chickpeas) and in a number of countries (Algeria, Egypt, Eritrea, Ethiopia, Iran, Jordan, Syria and Yemen), the role of the scientists is to make the crosses (mostly between landraces and between improved cultivars and landraces and wild relatives), grow the first two generations of crops on research stations, assess traits the farmers have defined as important, analyze the data and keep a safely stored electronic copy of the information. The farmers routinely evaluate and score the breeding material, decide what to maintain and what to discard, adopt and name varieties, and produce and distribute seed of the adopted varieties.

The testing process occurs in four stages: initial yield trials, advanced trials, elite trials and large-scale trials. The initial yield trials in Syria included 165 varieties. When crop diversity is great and farmers in different villages have different preferences, the initial trials in the villages use different varieties and only a few (usually five) common checks (traditional varieties used by local farmers). In these cases, the total number of varieties tested can be fairly large: in Syria, more than 400 genetically different varieties were tested. As there is only one initial trial per village, choosing which farmer will be involved and which field will be used is a serious decision requiring careful discussion with the farmers. If an unfortunate choice is made, for example, conducting the trial in the field of a farmer who is using agronomic practices different from those of most other farmers in the village, the resulting selections may not be well suited to the rest of the village.

The advanced and elite trials, which test the varieties selected during initial and advanced trials of the previous year, include two replications. Statistical analysis of the data is used to produce the best linear unbiased predictors of genotypic values and a number of variables including heritability. The large-scale trials use a replicated block design with very large plots and farmers' fields as the replications. Thus, the PPB trials generate the same quantity and quality of data as those obtained from multi-environment trials used in conventional breeding programs. In addition, they provide information about farmers' preferences, which is not usually available from conventional trials. Because the data are so sound, the resulting varieties usually qualify for official release. In several countries, including many in the developing world, this is a prerequisite for commercial seed production dictated by law or ministerial regulations.

Increasing crop diversity

One key aspect of this PPB model is that, once it is fully implemented, the lines selected as best are used as parents in a new cycle of recombination and selection, just as in a conventional breeding program. The difference is that these lines have been selected by farmers and can vary from location to location. This cyclic aspect, where farmers' best selection is used to produce the following generation, has an enormously empowering effect on the farmers, who feel their choices are valued by the breeder, and creates a strong sense of ownership among them.

In this PPB model, particular care was taken to design a scientifically robust model for two reasons. First, the farmers could be provided with scientifically correct information (the same type of information a breeder usually has) on which to base their decisions. Second, PPB programs are often criticized, sometimes rightly so, for not using a rigorous experimental design or statistical analysis; this model can withstand such criticism.

Because of the decentralized selection process and farmer participation, the PPB process leads to increased crop biodiversity. The number of different varieties at the end of a breeding cycle in farmers' fields is greater than the number of lines the Syrian National Program uses in its on-farm testing, which occasionally results in only one or two recommended varieties across the country. Many more varieties are adopted in the PPB program. This increase in biodiversity takes place not only in space (because different villages select different lines) but also in time, because of the cyclic nature of the process, which ensures rapid turnover of variety at the same location.

On average more than 1,000 farmers benefit from the program each cycle. During the second phase, the number of farmers directly involved in the program varies from 5 to 10 per village at the time of selection and from 10 to 15 per village at the time of data discussion. As a result, 200–400 farmers are directly involved in two of the most important decisions during each cropping season. In addition, in some villages, as many as 60 farmers buy seeds of the varieties selected through the PPB program.

A number of farmers have started to produce seed from the resulting PPB varieties. Because they are buying seed of a variety they have seen grown in the field by a farmer whose agronomic practices are similar to their own, farmers are sometimes willing to pay more than they would for little-known "improved" varieties available on the market. They also usually buy small amounts (100–200 kg) of seed because they subsequently multiply it. Therefore, the buyers in turn become seed producers and the benefits derived from the new varieties spread.

Everyone gains

As PPB progressed, farmers also contributed by suggesting changes in methods. In the beginning, visual selection occurred in the field, as requested by the farmers, on a day close to harvest time. That day, the farmers would gather, a short explanation would be provided for newcomers and each farmer would be given a score sheet for each trial. The farmers would then score each plot. At some locations, this could take up to half a day, at the end of which the scientists would collect the score sheets to enter the data into their computer programs. Visitors interested in the project would often be invited to these gatherings.

In 2005, Majid Awad, a farmer from Bylounan in Raqqa province, one of the driest villages taking part in the project, declared that he was not happy with this procedure. He complained that he could not concentrate properly on the scoring, a process he regarded as very central to future selection, because of frequent interruptions by visitors asking questions and walking in front of him as he was rating crops. He also pointed out that even though the selection day was chosen in consultation with the farmers, a last-minute commitment could prevent a farmer from attending and thus cause him or her to lose the opportunity to participate in the selection.

He suggested that the score sheet be distributed to all interested farmers well ahead of time, giving them the opportunity to choose when to do the scoring. They would be able to take as much time as they needed and even repeat the scoring if a climatic event changed growing conditions. (This had occurred one year when the various lines reacted differently to a heat wave after the selection day, and the farmers decided to repeat the scoring process.) The system Awad suggested was eventually adopted by the other villages, even though most of the farmers still preferred to set aside one day to discuss various aspects of the trials with the scientists.

Another modification of the method was related to the use of mixtures. Given that farmers in Syria do not generally plant heterogeneous plots, the ICARDA scientists were surprised to learn that Abdu Sheiko had decided to mix two very different barley varieties: a two-row variety, susceptible to lodging but drought resistant, and a six-row, lodging-resistant variety that produced a high yield in years of heavy rainfall. He explained that he had learned about the characteristics of the two varieties by conducting PPB trials and taking notes, and thought that mixing them could be a good strategy to stabilize yields. When other farmers were told about Abdu Sheiko's mixtures, some of them began mixing their leftover seed after samples had been taken to measure the yield. In the last three years, these mixtures have been producing better yields than any single variety; thus, the scientists and farmers decided to include experimental mixtures as part of the testing. This, in turn, contributed to the development of a program on evolutionary PPB because the farmers accepted the idea that mixtures can change with time in the direction of better-adapted genotypes.

Evolutionary PPB uses broadly diversified germplasm and long-term natural selection processes in the relevant areas to produce highly adapted crops. It also allows some degree of adaptation of the genetic material and increases the capacity of local communities to manage their seed populations. The handling of complex populations is very simple as all that is needed is to cultivate them in locations affected by either abiotic or biotic stresses or both, and let natural selection slowly increase the frequency of the best adapted genotypes. With the experience and skills they have developed through PPB, farmers and breeders can superimpose artificial selection for traits that are important at each specific location. Different farmers may select different plants and grow the progenies in their own field over many years; the expectation is that the varieties derived from this evolving population will be better adapted than those of preceding years.

These two examples show that farmers take the projects seriously and have ideas about how they can be improved. Farmers' experience should be taken into account and their suggestions incorporated into PPB projects. The degree to which information spreads from farmer to farmer and village to village also demonstrates how farmers learn from each other and experiment with new methods they think might be beneficial.

In 2010, to facilitate the sharing of lessons learned among the farmers, five computers were distributed to PPB participants in five villages. Farmers had expressed an interest in enhanced communication with ICARDA scientists and with other farmers participating in the program and in accessing information about agronomic management available online. The computers will also be used for the discussion of results of the PPB trials in farmers' fields.

The gender dimension

In 2006, a study revealed that women farmers in Syria were interested in PPB but were not being informed about the possibility of collaborating or were assuming they could not participate. Since then, a female researcher has been supporting the integration of Syrian women farmers into the PPB efforts by combining gender analysis with action research.

Participatory fieldwork has revealed gender-based differences in agronomic management, crop preferences and needs. Multi-criteria mapping was used to determine women's expectations of the program, their views on the validity of the current PPB process and their suggestions for improvement. PPB activities are now organized in ways that facilitate the involvement of women farmers by organizing events directly with women as well as collaborating with local institutions and creating women-only venues. The team tries to respect local sensitivities, particularly with regard to the participation of young female farmers in public events, and to create arenas for discussion that make it easier for women to interact with male strangers.

ICARDA also feels that it is important to create opportunities for women, men and ICARDA staff to collaborate, and it organizes mixed meetings and opportunities for sharing common concerns and implementing solutions. PPB activities are evaluated along with the farmers to gain a gender perspective on any problems that have been encountered. Gender issues are also taken into account when it comes to knowledge sharing. Because women, on average, are more illiterate than men and have less access to technology, reports are produced in both digital and hard copy, and include visual and oral material. In addition to these changes in approach and methods, the PPB project is expanding to include crops other than barley—e.g. chickpea and cumin—to reflect women's priorities and including priority traits for selection that were suggested by women—e.g. spike hardness, which is necessary for hand harvesting and palatability, and stem flexibility, which is important for handicrafts.

PPB, therefore, can accommodate varieties relevant to both women and men farmers who are often involved in complementary agronomic activities that entail different priorities and knowledge. Moreover, PPB facilitates access by women farmers to good seed supplies and information. This is a key element in the empowerment of women farmers in Syria who are generally disadvantaged in terms of access to resources, revenue and information. A study on the gender aspects of seed governance and PPB in Syria is currently underway.

A key challenge to achieving gender-balanced PPB in a patriarchal country, such as Syria, is ensuring that the participation of women farmers is an empowering and enriching opportunity for them, their households and communities. When this is achieved, the participation of women in public events is likely to be supported rather than resisted by their communities, and the benefits of the program can be shared more equally between men and women.

Benefit sharing

Data from the last few years, including the very dry 2008, show that the PPB lines outperformed both the commonly used landraces and conventionally bred modern varieties. In Kherbet El Dieb, which received rainfall of 189.5 mm in 2006, 206 mm in 2007 and only 139 mm in 2008, four PPB lines outyielded the local black-seeded landrace grown by most farmers by 12.3–23.2%. During visual selection, Al-Hassoun and the other farmers also scored the four lines higher than the landrace. The farmers from Kherbet El Dieb estimate that, in 2009, about 5,000 ha of the cultivated land in the area were planted with varieties introduced through the PPB program four years ago, then multiplied by the farmers. In 2010, they estimate, 90% of the farmers in the area planted one of three PPB varieties selected in the last five years. This estimate, which is based on the amount of seed sold and distributed, illustrates how successful the project has been in terms of variety adoption.

In Om El Amad, a village in the province of Hama with an average annual rainfall of 249 mm in the last four years (range: 183 mm in 2008 to 328 mm in 2007), the two best lines outyielded the local white-seeded landrace by 11–19% and a conventionally bred modern variety by 5–13%. In Bari Sharky, a drier village in the same province with an average annual rainfall over the last four years of 204 mm (range: 130 mm in 2008 to 238 mm in 2005), the largest yield increases were obtained with two lines resulting from crosses with the wild progenitor of barley. These lines outyielded the local landrace by about 33%.

But the selected lines are superior not only in marginal and drought-affected areas. In Suran, another village in Hama province, average annual rainfall over the last four years has been 277 mm. In three of these years, it received more than 300 mm; in 2008, it received only 198 mm. In this area, two sister lines obtained from crosses with landraces outyielded the local landrace by 15–25% and a conventionally bred modern variety by 18–27%.

All these lines are currently grown by farmers in the four villages and the seed will be distributed to other farmers. According to Ali Turkia from Tel-Hassan Bash, everyone who saw how the "Yana mixture" (a mixture of seed from the advanced, elite and extended trials in his field) grew requested seed for the next season as they were impressed with the plant height and spike length of the new variety, in particular. Compared with the local barley variety in this area as well as the conventionally bred Furat 2, the mixture performed very well.

Thus yields can increase and livelihoods can be improved by farmer participation in the breeding process. PPB studies in Syria have shown that no matter how many varieties are released and how much higher their yields are than local varieties, farmers in marginal environments will not adopt them unless they have participated in their selection. This makes PPB a particularly important tool in benefit sharing. Cost–benefit analysis of barley production at the farm level shows that participation of farmers in the breeding program does not mean higher costs of production. Farmers adopting varieties bred through PPB projects would likely pay higher input costs, but gain higher net returns. In addition to the economic returns, participating farmers appreciate other benefits, such as increased knowledge of barley production and variety selection and collaboration with scientists and other farmers. This demonstrates the importance of PPB and farmer participation. The benefits for women farmers, in particular, highlight the importance of adopting a gender-sensitive approach.

Cost–benefit analysis

The economic benefits of PPB are clear. Cost–benefit analysis showed that there is more to gain by implementing PPB than by continuing conventional plant breeding. Market-level benefits, calculated from the estimated adoption rate and yield gain, were compared with investment costs for PPB and conventional plant breeding. Even assuming only a 10% adoption rate and a 33% gain in yield for the varieties produced in the PPB programs, the benefit–cost ratio, as well as the internal rate of return, was higher for PPB crops. Because the impact of PPB depends on the availability of seeds from the resulting varieties, it is important to ensure that farmers, especially those on marginal lands, have access to these seeds.

The farmers benefited in other ways as well. The knowledge they gained through their participation in the program has improved their ability to make decisions regarding variety testing, evaluation and selection. Almost all the participating farmers say that, even if the PPB process ends, they will continue to practice what they have learned about variety selection. They also intend to maintain seeds of the new varieties and keep looking for good varieties along with other farmers. Many feel that their participation has improved their knowledge of barley production, as well as agriculture in general.

Working with researchers is assumed to improve the "human capital" of participating farmers, and some, women in particular, did feel that their knowledge has increased as a result of their interaction with breeders and technicians. The women farmers also believed that their role in agronomic management, usually overlooked at household and village levels, and by researchers and development practitioners, had become more visible through PPB.

Working in groups and being encouraged to share information and knowledge may lead to increased "social capital," in terms of ability to cooperate and share information. Many of the participating farmers said that they gained valuable experience through interactions with other farmers. One of the most important successes of the PPB program was that it had a positive impact on the livelihoods of most of the participating farmers. Most farmers who have not yet felt the impact on their livelihoods live in areas where the PPB program started later, and it is likely that their situations will improve as PPB continues. Women farmers particularly valued their increased access to good seed and information.

Only a very small number of farmers believed that those who were involved in selecting new varieties should keep the benefits for themselves; most felt that the benefits should be shared at the community level. This might indicate that the farmers view local plant genetic resources as their common heritage, not something only a few should benefit from. Other projects will probably also be more in tune with the values of the farming communities if they take cooperation, sharing and equal distribution of benefits as their point of departure.

It is commonly thought that Syria's legislation regulating variety release and seed multiplication and distribution has been an obstacle to the participatory barley breeding project by limiting the amount of seed that can be produced and distributed, thus preventing thousands of farmers from benefiting from the project. However, the only legislation in this area is a Ministerial Decree from 1975 (available only in Arabic), and it does not contain any specific restrictions on the movement of seed. The legislative situation with regard to this issue may be somewhat unclear, and the uncertainty surrounding the legality of seed distribution might be a barrier to upscaling. The Ministry of Agriculture and Agrarian Reform is currently in the process of drafting a seed law, including a new system for releasing new varieties. This law will probably bring legal certainty to the field, but if it places restrictions on the exchange of seed, it might also be detrimental to farmers' rights.

Conclusions

The participatory program in Syria has already inspired other countries in the region (Algeria, Egypt, Eritrea, Ethiopia, Iran, Jordan, Morocco, Tunisia and Yemen) to start PPB of several crops. One of the most important lessons for those seeking to copy this project's success is that similar projects should also start by involving their national institutions with responsibility for plant breeding. It can be argued that only by institutionalizing PPB can the method achieve full impact. To ensure the success of such projects, especially in reaching out to a substantial number of farmers, it is also crucial that seed laws allow the necessary seed multiplication and distribution.

PPB gives farmers the opportunity to influence the development of technologies that are better adapted to their specific needs, agro-ecological environments and cultural preferences. It also provides them with the opportunity to influence decisions about how financial resources for research and agricultural extension services are used. In addition, the project makes use of the traditional knowledge of farmers and, thereby, elevates the profile of that knowledge and its holders, creating incentives to continue using and developing it. Although PPB is still not a very widespread practice, it can be structured to provide opportunities for women to contribute to the development of varieties relevant to the food chain and to enjoy the benefits of PPB. That is what the project in Syria has tried to do with its gender-sensitive approach.

Participatory processes also bring farmers into contact with professional breeders, making the farmers more aware of what science can offer them. This awareness can have an empowering effect, something that can be seen in the enhanced quality of the Syrian farmers' participation over time as they become true research partners. The farmers are involved not only in breeding activities, but also in the registration of the resulting varieties, their maintenance, seed multiplication and distribution and, as appropriate, commercialization. PPB has also strengthened the seed systems by improving production, selection and access to seeds. Along with increased yields, this is an important contribution to food security in Kherbet El Dieb and the other villages involved.

Finally, it is worth mentioning that by increasing access to better-adapted and higher-yielding varieties, PPB can contribute to ensuring the right to food. In fact, PPB is one of the recommendations of the interim report of the Special Rapporteur on the right to food, who also places special emphasis on the importance of collaborating with small-scale, women and marginal farmers (United Nations Special Rapporteur on the right to food 2010).

References

United Nations Special Rapporteur on the right to food (2010) Mission to Syria from 29 August to 7 September 2010. Office of the United Nations High Commissioner for Human Rights, Geneva, Switzerland. Available at: www.srfood.org/images/stories/pdf/officialreports/20100907_syria-mission-preliminary-conclusions_en.pdf (accessed 10 February 2011).

WFP (World Food Programme) (2011) World hunger. WFP, Rome, Italy. Available at: www.wfp.org/hunger (accessed 10 February 2011).

7 Jordan

In search of new benefit-sharing practices through participatory plant breeding

Adnan Al-Yassin

Supporting agricultural development

Of Jordan's 8.93 million hectares, only 7.8% is arable land. However, the agricultural sector plays an important role through its contribution to national income and employment. Agriculture contributes 7.5% to gross national product, and about 22% of Jordan's population (estimated at 5,835,500) make a living from agriculture.

The country has a Mediterranean-type climate and several agro-climatic zones, which vary considerably in terms of rainfall, temperature, soils and cropping patterns. Agricultural crops are mainly rainfed (98%). Field crops (such as cereals, food and feed legumes), orchards (mainly olive trees) and vegetables are grown on 65.5%, 25.5% and 9.0%, respectively, of agricultural lands. Wheat, barley, lentils, chickpeas and vetches are produced during the main winter growing season. Irrigated agriculture is concentrated in the rift valley (the Jordan Valley), where vegetables and citrus fruits are the main crops. In the southeastern part of the country, cereals and forage crops are grown using pivot irrigation. Sources of water are the Jordan River, springs, wells and several dams.

Jordan's government has been active in creating a supportive institutional environment for agricultural development. This case study looks at agricultural policies, laws and international agreements through the lens of the country's efforts to introduce and institutionalize PPB in collaboration with ICARDA. These PPB activities build on ICARDA's pioneering work in Syria and other countries (see Chapter 6).

ABS issues are still new to the country, but are attracting attention. The ABS team, which is made up of staff from the National Center for Agricultural Research and Extension (NCARE) and ICARDA and is part of the IDRC-supported project on ABS issues, is at the forefront of efforts to gain more recognition.

Policies and laws in the agricultural sector

Government policies support development of the agricultural sector by expanding the area under cultivation and improving the supply of inputs. The government also encourages new technology and crops by: employing better approaches to research and extension, rainwater harvesting techniques and irrigation systems;

controlling input prices; promoting agricultural development projects; and supporting guaranteed minimum prices. The government buys local wheat and barley at international prices to encourage farmers to increase production and helps farmers export their surpluses of other crops.

At the regional and international levels, the government has ratified the following international treaties and conventions regarding biodiversity and the environment:

- Ramsar Convention in 1977
- Convention on Biological Diversity (CBD) in 1993
- Convention to Combat Desertification in 1996
- Cartagena Protocol in 2000
- Kyoto Protocol in 2000
- Plant Genetic Resources for Food and Agriculture in 2001
- Convention on Persistent Organic Pollutants in 2002
- International Treaty on Plant Genetic Resources for Food and Agriculture (ITPGRFA) in 2004
- World Heritage Convention
- Regional Convention for the Conservation of the Red Sea and the Gulf of Aden Environment.

In addition, Jordan adopted the Standard Material Transfer Agreement of the governing body of the ITPGRFA in its resolution 1/2006 on 16 June 2006. In October 2004, it became a member of the International Union for the Protection of New Varieties of Plants (UPOV).

National legislation

Although Jordan is strongly aware of environmental and pollution issues, it still has relatively limited knowledge of the importance of plant genetic resources. Jordanian society learns quickly, however, and the establishment of the Genetic Resources Unit and the National Committee will play a major role in increasing awareness in this area.

Quarantine laws in Jordan are not strict enough to inhibit the transfer of genetic materials. The country is freely receiving germplasm, mainly cereals, from international research centers, such as ICARDA. The flow of germplasm abroad is usually not checked either. The Genetic Resource Unit and the Ministry of Agriculture are expected to play key roles in controlling these flows of genetic resources.

Jordan's program for producing and certifying cereal seeds is going well; however, help is needed for variety release, as much effort is going into breeding new varieties from landraces or introduced germplasm.

Existing regulations governing the import and export of seeds and agricultural produce include those on: control of seed production (1987); conditions for variety registration (1990); conditions for seed trade (1990); seed trade of agricultural crops (1990); licensing seed companies (1990); licensing agricultural companies for seed import (1990); variety registration of agricultural crops (1993); seed production and trade of cereals, forages, vegetables and fruit trees (1996).

Under the agricultural law, seeds and plants imported for multiplication are exempt from taxes. For example, the private sector is allowed to import inbred lines, tax free, to encourage seed production locally. In 2000, the government enacted Law 24 for the protection of new plant varieties, which takes into account WTO and UPOV agreements and conventions. This law describes the requirements for protection of "new plant varieties" and covers other related legal issues, such as right of priority, provisional protection, publication, licensing and ownership, cancellation of registration, general rules and variety denomination. The four essential conditions for obtaining rights to a variety under this law are distinctness, uniformity, stability and novelty. No reference was made to PPB varieties in Law 24. The Ministry of Agriculture was responsible for preparing related regulations and directions and implementing them in the second half of 2002.

Jordan attended the United Nations Conference on Environment and Development in Rio de Janeiro, Brazil, in 1992, and was part of the discussion surrounding Agenda 21. On 5 June 1992, Jordan signed the agreement on Biodiversity and Climate Change, and, on 9 November 1993, a royal decree was issued to approve implementation of this agreement. This helped Jordan move forward on many aspects of the management of biodiversity, including plant genetic resources; for example, the government formed a national committee for the conservation of biological biodiversity that included representatives from the Ministry of Agriculture, the Ministry of Planning, the Ministry of Tourism and Antiquities, the universities, the Royal Society for the Conservation of Nature, the Society for Protection of the Environment and the Department of Environment.

Researchers and policy analysts realize that one of the major causes of agrobiodiversity degradation is inappropriate legislation and policies. Alternative options are needed in several areas. For example, domestication of international agreements and conventions and harmonization of regional policies and legislation affecting the conservation of agrobiodiversity would ensure a coherent regional approach to addressing some of the legal issues surrounding biodiversity conservation. The initiative of the African Union in developing Model Legislation on the Protection of the Rights of Local Communities, Farmers and Breeders and for the Regulation of Access to Biological Resources could serve as a model. Discussion of this idea continues.

Agriculture Law 44 (2002) and its guidelines cover variety registration, seed production, seed processing, seed marketing, seed quality control and seed trade (import–export). Multiplication, production, processing and marketing of the seeds of any cultivar are prohibited unless the cultivar is registered as described by the law.

Neither Law 24 nor Law 44 takes into consideration the vital role of farmers; they provide protection only to crop varieties developed through conventional methods. A model law is needed that includes the concept of farmers' rights, although developing such a law will require considerable time and effort. Similar models have been developed by many African nations and India to focus attention on the conservation and sustainable use of biodiversity, food security, protection of community rights (including farmers and breeders), equitable sharing of benefits consistent with the provisions of CBD and the concept of national sovereignty.

Plant breeding research

Jordan is endowed with a wealth of genetic resources, both cultivated crops (barley, wheat, lentils, chickpeas, figs, olives and capers) and wild relatives (particularly barley, wheat, lentils, chickpeas and pistachios). These genetic resources are available from the National Gene Bank (which holds approximately 5,000 accessions) and ICARDA's gene bank (with more than 132,000 accessions, representing over 20% of the world collection held in trust by the CGIAR centers), as well as in situ. Because of the harsh environment, conventional plant breeding has not produced varieties to replace the landraces of the main field crops, with the possible exception of wheat.

NCARE, the country's leading agricultural research agency, has had the task of managing breeding programs in cereal grains since the 1950s. The major output of this ongoing program was the release of six barley varieties, 12 wheat varieties, and one *Vicia* variety. In addition, three chickpea and three lentil varieties were the result of collaboration between NCARE and the University of Jordan in the 1980s. Unfortunately, no new varieties appeared subsequently until 2004, when NCARE submitted three barley varieties and two wheat varieties to the Varieties Release Committee, which is chaired, according to statute, by the director general of NCARE.

Jordan's introduction to PPB took place in 2000 when it engaged in an IDRC-supported ICARDA project entitled From Formal to Participatory Plant Breeding: Improving Barley Production in the Rainfed Areas of Jordan (2000–03). NCARE was a partner in this project and implemented activities in farmers' fields. This established a new direction for the national breeding program: from centralized to decentralized breeding work. The logic of a decentralized approached is illustrated in Figure 7.1. Conventional plant breeding is a cyclical process that takes place largely at one or more research stations with the breeder making all decisions. PPB is the same process, but it takes place mainly in farmers' fields and decisions are made jointly by farmers and breeders.

Introducing PPB in Jordan resulted in a dramatic change in the attitude and behavior of breeders. They came to acknowledge and appreciate the knowledge and skills of farmers (both women and men), and began to look for ways to build on their expertise. They also became aware that benefits include not only the final products of the breeding process (i.e. improved and released varieties), but also the sharing of knowledge and experience, which led to new insights, new experiences, new diversity and a step-wise improvement in farmers' crops and seeds. This was a major discovery and an important step in opening up the conventional approach and system.

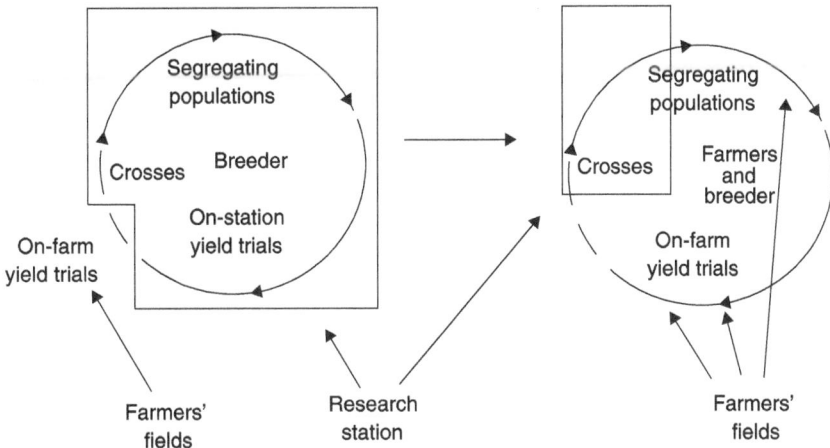

Figure 7.1 Comparison of conventional and participatory plant breeding
Source: Ceccarelli and Grando (2007)

NCARE's objectives in undertaking this project were to promote PPB in Jordan, to improve barley varieties, to enhance the rate of adoption of new varieties through farmer participation in selection and testing, to identify differences between selection criteria used by men and women farmers and by breeders, and to disseminate experimental results.

During the first three years, good results were obtained and the objectives were achieved to a large extent. However, little or no progress was made in terms of the policy and legal implications of these efforts, although the research team spent considerable energy creating awareness of the new breeding approach (among the research and policy communities) and how it could be adapted to and benefit the country. Farmers who took an active part in the research were happy with the results. When the initial project ended, they called on ICARDA and NCARE to continue with the PPB process and expand on it.

The farmers' voices were heard and respected. ICARDA took the lead in developing and implementing a follow-up project entitled Institutionalizing Participatory Plant Breeding within National Plant Breeding Systems: Costs and Benefits of Seed Production (2004–07), also funded by IDRC. It was during implementation of this second initiative that questions of ABS became more central. The team not only continued to improve and expand PPB work in the field, but also aimed to achieve better understanding of the constraints on PPB related to variety release, certified seed production and intellectual property rights. For the first time, the team acknowledged the importance of farmers' rights. Farmers themselves also began to realize that there are important policy and legal issues related to PPB, although they are not always and immediately visible to them.

In 2005, the voices of farmers were heard again when they called on ICARDA and NCARE to implement PPB for other important crops, especially wheat and chickpeas. This new PPB work started in 2005–06, at a time when PPB practices

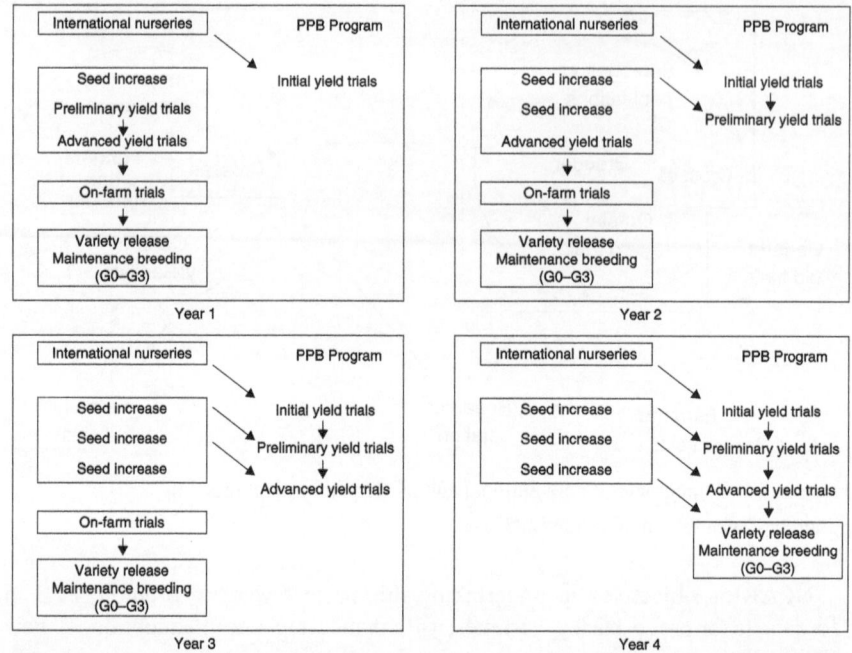

Figure 7.2 Transformation of NCARE's conventional plant breeding system into a
PPB system

Source: Al-Yassin (2005)

were being integrated into university curricula with input from seed specialists.
The aim was not only to broaden the scientific base for PPB, but also to build up
strong evidence for policymakers that the science was backed by practice, which
would lead to better adoption and adaptation. ICARDA and NCARE appealed to
the inter-country Consultative Group for Participatory Plant Breeding for support
to influence the policy agenda further. In May 2005, ICARDA held an important
meeting with the consultative group, in which NCARE expressed its intention to
modify the entire breeding program for all cereal grain crops and use a PPB
approach, marking the beginning of the institutionalization process. The step-wise
procedure it proposed is shown in Figure 7.2.

PPB methods

The PPB model consists of several stages (Figure 7.3). Farmer initial trials (FITs)
were conducted to measure yields of early segregated populations. These were
unreplicated trials on 200 plots of 12 m^2 encompassing 170 varieties plus controls
(one or two controls repeated 30 times). Breeding material selected from the FITs
was tested the second year in farmer advanced trials (FATs), with the number of
varieties and controls varying from village to village. For the FATs, plot size was
45 m^2 to produce sufficient quantities of the selected seeds to be planted on larger

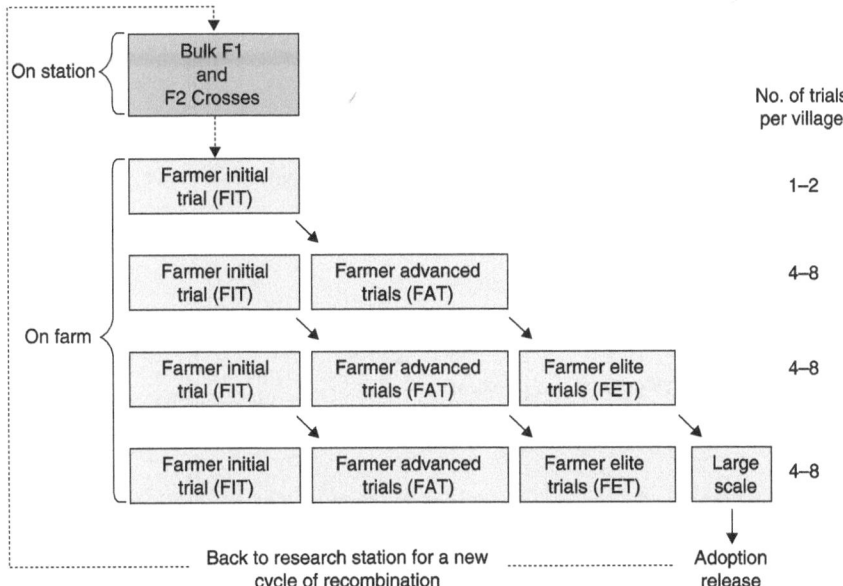

Figure 7.3 Model of the PPB system organized with farmers in Jordan

Source: ICARDA (2007)

plots in the third stage. The number of FATs in each village depended on how many farmers were willing to engage in this type of trial. In a given village, the FATs evaluated the same varieties, regardless of the number of farmers. Each farmer decided on the rotation, seeding rate, soil type, the amount of fertilizer used and the timing of application. Thus, the FATs took place under a variety of field conditions and management systems. During selection, farmers exchanged information about agronomic management, and relied greatly on this information in deciding which varieties to select. Thus, favouring characteristics of the crops in terms of their response to environmental or agronomic factors started at an early stage of the selection process.

The model in Figure 7.4 shows how formal and informal seed systems are integrated in the PPB process. During selection and testing, i.e. the FITs, FATs and farmer elite trials (FETs), which represent a gradual scaling-out sequence, the required amount of seed of each variety usually varies from 50 to 100 kg. Likewise, the number of varieties planted in each village ranges from 15 to 30. In the conventional seed system, varieties are produced, cleaned and treated on station. Now, the objective is to have these processes take place in villages using locally manufactured seed cleaners. These cleaners should include a device to treat the seed to make it disease resistant and they must be able to process about 400 kg of seed an hour. The community-based seed multiplication system is a model for informal seed dissemination and also represents a concrete way to improve local access to clean seeds and generate benefits for local people.

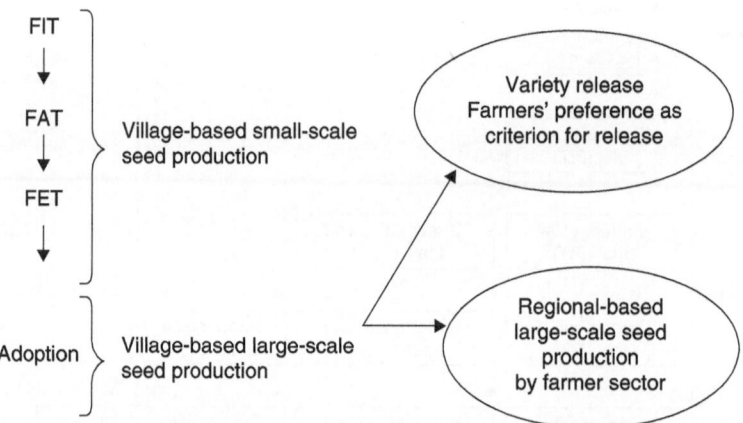

Figure 7.4 Linking PPB and variety release with informal and formal seed production
Source: Ceccarelli and Grando (2007)

During the PPB trials, both the national conventional breeding program and the PPB program were operating. No new varieties of wheat or chickpeas have reached the scale-out phase yet, because of frequent droughts. However, barley PPB varieties have now reached the farmers' multiplication phase. The key issues that have emerged are seed handling, legislation and benefit sharing.

In 2004, NCARE submitted three barley varieties and two wheat varieties, produced from the conventional breeding program, to the Varieties Release Committee. The released barley varieties are Athroh (six-row), Yarmouk (two-row, Esp/1808-4L//Harmal) and Muta'a (two-row, Roho/A.Abiad/6250); the wheat varieties are Ammon (BW, Tsi/vee's') and Um Qais (DW, Om rabi5). Unfortunately, only Um Qais has reached farmers' fields. The others are still at the multiplication stage, because of frequent crop failures resulting from drought.

However, about 30 barley PPB varieties were scaled out to six regions. These varieties are adapted to diverse conditions found in farmers' fields and are in demand as they respond to the interests and needs of larger farmers, seed growers, sheep owners, combine harvesters, farmers in low-input agriculture and women farmers doing handicrafts. Some of these varieties are shown in Table 7.1.

Policy and legal context for ABS in Jordan

The evolution of the PPB process led almost naturally to the realization by the research team that the breeding programs were not just a matter of technical expertise, but that important policy and legal issues also have an impact on PPB. The success of the barley program and the expansion to other crops created a need to address these issues, reinforced by growing international awareness and pressure to deal with them.

Table 7.1 Farmer-adopted PPB barley varieties with specific adaptation to diverse conditions found in farmers' fields

Region	Nr	Name
Ramtha E	1	Arar/Lignee527//Arar/PI386540
	2	Moroc9-75//WI2291/CI01387/3/H.spont.41-1/Tadmor
	3	ArabiAbiad/Arar//H.spont.41-5/Tadmor/3/H.spont.41-1/Tadmor
	4	Arar/H.spont.19-15//Hml/3/H.spont.41-1/Tadmor/4/WI2291/Tadmor
	5	Moroc9-75//WI2291/WI2269
	6	Roho/4/Zanbaka/3/ER/Apm//Lignee131/5/Akrash//WI2291/WI2269/3/WI2291/WI2269//WI2291/Bgs
Ramtha W	1	Alanda/3/CI08887/CI05761//Lignee640/4/Alanda/Lossaika
	2	Cerise/Lignee1479//Moroc9-75/PmB/3/JLB37-74/H.spont.41-5//JLB37-74/H.spont.41-5
	3	ArabiAbiad/Arar//H.spont.41-5/Tadmor
	4	ChiCm/An57//Albert/3/Alger/Ceres362-1-1/4/Arta
	5	Soufara-02/3/RM1508/Por//WI2269/4/Hml-02/ArabiAbiad//ER/Apm
	6	Roho/4/Zanbaka/3/ER/Apm//Lignee131/5/WI2291/Tadmor//Arta
Khanasri	1	Kv//Alger/Ceres.362-1-1/3/WI2269/4/Sara
	2	Moroc9-75//WI2291/CI01387/3/H.spont.41-1/Tadmor
	3	WI3159/5/Roho/4/Zanbaka/3/ER/Apm//Lignee131
	4	Roho/4/Zanbaka/3/ER/Apm//Lignee131/5/Arta
	5	Arta//Moroc9-75/ArabiAswad/3/WI2291/Tadmor//Arta
	6	ArabiAbiad/Arar//H.spont.41-5/Tadmor
	7	ChiCm/An57//Albert/3/Alger/Ceres362-1-1/4/Arta
Ghweir	1	ChiCm/An57//Albert/3/Alger/Ceres362-1-1/4/Arta
	2	Zanbaka/5/Pyo/Cam//Avt/RM1508/3/Pon/4/Mona/Ben//Cam/6/Sara
	3	WI3167/4/Arta/3/Hml-02//Esp/1808-4L
	4	Arta//Moroc9-75/ArabiAswad/4/Akrash//WI2291/WI2269/3/WI2291/WI2269//WI2291/Bgs
	5	Sara/4/H.Spont.96-3/3/Roho//Alger/Ceres.362-1-1
Rabbah	1	Arta//Moroc9-75/ArabiAswad/6/WI2291/4/7028/2759/3/69-82//Ds/Apro/5/Zanbaka/3/ER/Apm//Lignee131
	2	WI3277/4/Arta/3/Hml-02//Esp/1808-4L
	3	WI3159/5/Roho/4/Zanbaka/3/ER/Apm//Lignee131
	4	Sara/4/H.Spont.96-3/3/Roho//Alger/Ceres.362-1-1
	5	Zanbaka/5/Pyo/Cam//Avt/RM1508/3/Pon/4/Mona/Ben//Cam/6/Arta
Mohai	1	ChiCm/An57//Albert/3/Alger/Ceres.362-1-1/4/Arta
	2	Limon/Bichy2000/5/Roho/4/Zanbaka/3/ER/Apm//Lignee131

Source: Compiled by the author.

Thus, the team began to discuss policies and laws related to genetic resources with farmers. Many farmers chose to be represented by the farmers' union in fora where policies and laws related to benefit sharing are formulated. A growing interest in questions concerning traditional knowledge and benefit sharing emerged among farmers during meetings organized by the ABS team on 3 and 4 March 2009. They made this interest known to policymakers in several ways. First, they contributed to the gene bank database by evaluating almost 50% of Jordanian landraces of both wheat and barley, which they valued properly. Second, during the 8th Conference of the General Union of Arab Peasants and Agricultural Cooperatives in Amman on 22 March 2009, they claimed rights to their traditional knowledge and shared benefits from the use of germplasm.

The farmers were encouraged to push forward on the ABS issue when a farmer patented an extraordinary grapevine variety and gave it his name, Mansour 2000, but then did not benefit from its development. Taking notice of this, the farmers' union called on the government to pass a new bylaw for benefit sharing, in line with Article 9 of the ITPGRFA. This brief story illustrates that farmers are no longer just passive "recipients" of new technologies, policies and laws.

In due time, these efforts generated a response. Recently, NCARE developed a draft proposal on intellectual property rights and ABS issues. The proposal is still waiting to be approved and, if all goes well, implemented. Unfortunately, neither farmers nor their representatives contributed to it.

Variety release

In Jordan, variety release is usually the responsibility of the Varieties Release Committee, which is appointed by the Minister of Agriculture and chaired by NCARE's director general. This committee makes decisions based on scientific reports prepared by breeders. The reports cover performance, agronomic characteristics, reaction to pests and diseases and quality characteristics of the new variety. The members of the Varieties Release Committee represent the national research centers, the universities, the Jordan Cooperative Corporation (JCC), the extension service, NGOs and the private sector. All these stakeholders are involved in either the implementation of PPB or teaching PPB principles and methods. The National Research Center is the body responsible for submitting candidate varieties to the Varieties Release Committee (through its Field Crops Department).

Currently, the Varieties Release Committee guidelines do not allow for consideration of farmers' opinions; thus, there are several cases of varieties that were released but never grown by farmers and varieties grown by farmers that have not been released. When any new variety is not adopted, the considerable investment made in its development brings no benefit. It has been shown that the economic cost to farmers of releasing an inferior genotype is much less than the economic cost of not releasing a superior genotype.

In 2007, the Varieties Release Committee took steps toward accepting data from PPB trials as the basis for variety release. On 14 May 2009, the minister of agriculture reinforced the committee's position by attending a national workshop and spending an entire day making field visits in the Maru area. This push has motivated the ABS project team (led by NCARE and ICARDA) to publish new guidelines for releasing PPB varieties in Jordan.

Jordan may benefit from a national law on farmers' rights, but so far it has not been feasible to define clear ABS principles, especially concerning seed multiplication and distribution. Farmers usually raise ABS issues at their meetings, such as recognizing the different levels of participation that affect ABS from developed varieties. However, the farmers have their own interpretation of benefit sharing. Some of them have multiplied seeds and taken the initiative of distributing half to other farmers in the PPB research area for free. Another farmer is selling his new variety to any farmer who asks for it and writes their names in a notebook to be able to track the seed diffusion process. Thus, more than one "model" for equitable benefit sharing of newly developed varieties is in the making in the informal seed multiplication system. How to translate this into adequate policy and legislation remains a challenge.

Seed multiplication: formal versus informal

The seed multiplication and dissemination system is the responsibility of the JCC. It deals only with the seeds of officially released varieties, starting with maintenance breeding (growing first-generation or breeder seeds) until sufficient quantities for large-scale commercialization have been produced. This work started in 1982 with 12 wheat varieties and three barley varieties. Soon, it was noticed that eight wheat varieties and one barley variety were not being used by farmers, and multiplication of those seeds was stopped. The production of chickpea and lentil seeds began in 1992, but was interrupted a few years later.

After harvest, seed lots are stored at JCC stations and official samples are retrieved and submitted to the Central Seed Testing Laboratory in NCARE for quality testing. Samples that meet national seed standards are cleaned, treated and stored for the next planting season. JCC storage facilities are located at seed processing centres (Table 7.2). Seed lots are sprayed with insecticides during storage to protect them from infestation by pests. About 25% of cleaned and treated seeds are stored as surplus to supplement shortages during drought years. The surplus seeds are renewed regularly to maintain their vigour.

Despite the fact that the JCC is responsible for providing "certified" seeds of released varieties, about half of the land devoted to these crops is planted with uncertified seeds purchased from farmers. As this informal seed system is active,

Table 7.2 Location and capacity of seed storage facilities

Location	Crops	Capacity (t)	Annual supply (t)	Facilities
Ramtha	Wheat, barley	6,000	2,000	Warehouse
Mushaqar	Wheat, barley	4,000	2,000	Warehouse, silo

the ABS team aims to empower farmers to produce certified seeds, for example, by providing them with seed cleaners. Helping them to market PPB improved seeds will be a completely new way of generating benefits for farmers.

Conclusions

Key ABS issues emerged during the introduction, testing and upscaling of PPB. Although Jordan has adopted a comprehensive framework of agricultural policies and laws, ABS issues, especially in relation to PPB, have not yet been dealt with in a clear, concise, practical manner. The ABS team has made a start by identifying key issues in relation to the various elements of PPB, but the general lack of knowledge among researchers, policymakers and farmers has been a challenge.

Farmers speak out when they have the chance, and giving them such opportunities, through meetings, workshops and conferences, has pushed the PPB agenda forward. But farmers do not yet have any formal representation in important policy and legal fora. "Farmers' rights" is now a concept being discussed in the country, but whether it can be captured in legislation remains to be seen. Through trial and error, PPB research has created new ways to obtain access to genetic resources as well as new forms of benefit sharing, but, as yet, no clear guidelines exist for formal recognition in policies and laws. Farmers are trying out various ways to share benefits, some following more conventional practice, others more open to novel practice. The ABS team is working with these farmers, to document their practices and analyze the pros and cons of the various models for benefit sharing.

References

Al-Yassin, A. (2005) Presentation during the Consultative Workshop on Participatory Plant Breeding. International Center for Agricultural Research in Dry Areas, Aleppo, Syria.

Ceccarelli, S. and Grando, S. (2007) Decentralized participatory plant breeding: an example of demand-driven research. *Euphytica* 155: 349–60.

FAO (Food and Agriculture Organization of the United Nations) (2009) International treaty on plant genetic resources for food and agriculture. FAO, Rome, Italy. Available at: ftp:// ftp.fao.org/docrep/fao/011/i0510e/i0510e.pdf (accessed 1 April 2011).

ICARDA (International Center for Agricultural Research in Dry Areas) (2007) Participatory plant breeding: an example of demand-driven research (poster). ICARDA, Aleppo, Syria. ICARDA-03/1000/Revised/Feb.2007.

8   Honduras

Rights of farmers and breeders' rights in the new globalizing context

Sally Humphries, José Jiménez, Omar Gallardo,
Marvin Gomez, Fredy Sierra, and members of
the Association of Local Agricultural Research
Committees of Yorito, Victoria and Sulaco

Farmers' rights under pressure

In April 2006, Honduras entered into a free-trade agreement with the United States: the Dominican Republic–Central American Free Trade Agreement (DR-CAFTA). This agreement would have a profound effect on smallholder agriculture, on which a large percentage of the population in the poorest countries of Central America depends. In this case study, we examine the impact of this agreement on access and benefit sharing (ABS) with regard to the plant genetic resources of small Honduran farmers.

Although Honduras became a signatory to the Convention on Biological Diversity (CBD) in 1992, it was not until 2008 that it ratified the International Treaty on Plant Genetic Resources for Food and Agriculture (ITPGRFA), becoming party to that agreement in 2009. The ITPGRFA, which is in harmony with the CBD, addresses farmers' rights, which include the right to participate equitably in benefit sharing, protection of indigenous knowledge, the right to participate in decision-making at the national level regarding conservation of PGRFA and the right of farmers in poor countries to obtain the resources they need for food security (Cooper 2003: 470–4). The ITPGRFA provides signatories with rights that are contrary to the provisions of DR-CAFTA, and the Honduran government has delayed passage of the Law for the Protection of Plant Varieties since 2001, effectively stalling a decision on these conflicting rights.

The socioeconomic situation

Honduras is the third-poorest country in the western hemisphere after Haiti and Guatemala. It is also one with the greatest inequalities. Over the past decade, the income gap has increased and inequalities between urban and rural populations are the most visible. Although poverty is widespread in both rural and urban areas—69% of the rural population lives below the poverty line compared with 55% of the urban population—it is at the level of extreme poverty that differences are the most acute, with 60% of the rural population living in such conditions compared with 25% of the urban population.

Nearly half the population of Honduras lives in rural areas and depends on agriculture. However, agriculture accounted for only 23% of GDP in 2004 (World Food Programme 2005: 10). Averages mask the gap between the large agri-export sector (e.g. bananas, palm oil, pineapples, etc.) and small-scale producers of mostly staples (maize and beans). Small-scale agriculture is generally low yielding and aimed primarily at providing for the subsistence needs of household members. It is in this latter sector that the rural poor lie.

Income-earning opportunities, other than agriculture, in rural areas are more limited in Honduras than in other Central American countries. Limited non-farm income is linked to the bifurcated nature of agriculture in Honduras. Investment opportunities for small subsistence farms are minimal, while large mechanized farms demand little labor and, hence, place severe limits on the size of the salaried labor force, whose wages might otherwise contribute toward a more buoyant non-farm economy.

Unequal access to land and isolation from transportation networks and markets are primary factors behind rural poverty. Approximately 80% of Honduran farmers—some 400,000 households—own less than five hectares, totalling less than 15% of agricultural land. At the other extreme, 6,000 farms or holdings of more than 50 hectares occupy 30% of the total (World Food Programme 2005). A study by the World Bank (2004: 12) suggests that 1% of farmers hold 25% of the land. The largest of these properties are mainly located in the valleys and coastal plains of the north and northeast where the export sector predominates. While the larger properties are generally well connected to the central transportation network that forms a T across the north and down the centre of the country, small-farmer agriculture is largely outside this network, isolating poorer farmers from the country's markets and leaving them with land on remote hillsides that is often too steep for agriculture (Jansen *et al.* 2005).

The structure of the agricultural sector has also contributed to stagnant production of grains and pulses, the primary components of the national diet. Per capita consumption of white maize averages 80 kg a year and constitutes more than 30% of the caloric intake. Twenty-five to 30% of maize is produced on farms of under 2.5 hectares and an average of 40% is consumed within producing households; this increases to 65% among smallholders (World Food Programme 2005). Most small farms produce low yields. Marketable surpluses come from a handful of departments. However, even among commercial grain producers, yields have been stagnant over the past 20 years and the average has remained at around 1.46 t/ha (Centro de Desarrollo Humano 2005: 54). The performance of food producers has been negatively affected by the terms of trade faced by agriculture versus non-agricultural sectors. At the same time, since the early 1990s, Honduras has become increasingly dependent on food imports to feed its rapidly growing population. The impact of import dependence was most keenly felt in 2007–08, when steep rises in global grain prices severely affected poor households, whose incomes were largely devoted to food purchases, underlining the vulnerability of the majority of Hondurans to global integration.

By the turn of the 21st century, agricultural exports (both traditional and non-traditional) were no longer the country's major foreign income earners; rather the *maquila* sector (tax-free apparel assembly) and international remittances had taken over (Centro de Desarrollo Humano 2005: 49). DR-CAFTA supports the manufacturing industry by providing capital investors with the assurance of continued access to the United States, the world's largest market for consumer goods. The Government of Honduras hopes that remittances from the *maquila* sector as well as those from overseas will create a stimulus for non-farm activities within the rural sector and ameliorate rural poverty (Centro de Desarrollo Humano 2005: 48).

Plant-related treaties

With the introduction of DR-CAFTA, World Trade Organization rules took effect. These call for recognition of trade-related intellectual property rights (TRIPS), including those of the International Union for the Protection of New Varieties of Plants convention (UPOV-91). The latter effectively imposes a global legislative model of patent-like plant protection on all countries. Critics of the agreement argue that this serves the interests of seed companies and not small farmers (GRAIN 2003). Specifically, it gives multinational corporations full license to introduce genetically modified seeds into the region, fully protected under international patent agreements, while small farmers, whose seeds do not necessarily conform to the requirements for protection in terms of "distinctiveness, uniformity and stability," have no protection whatsoever. This leaves farmers open to exploitation through biopiracy or "bioprospecting," as foreign companies engaging in the practice prefer to call it (GRAIN 2006).

Just as farmers' seeds are unprotected under DR-CAFTA, so too is indigenous knowledge. The use of indigenous or traditional knowledge or local seed does not require prior informed consent, disclosure of origin or benefit sharing. This means that patent applicants do not have to provide explicit clearance for the use of these materials, nor is the origin of the material or an arrangement for benefit sharing between the applicant and the knowledge or seed holders required. Wording in the DR-CAFTA document limits "disclosure" for the purposes of patent protection to what is "sufficiently clear and complete." Indigenous knowledge can rarely be so defined. Thus, even though the document itself makes no reference to indigenous knowledge, the implication is that "failure to indicate the origin of a plant or show proof of consent for its use from a local community may never be grounds for rejecting a patent application" (GRAIN 2006). Simply put, the principles of disclosure of origin, prior informed consent and benefit sharing, drawn from the CBD and employed to prevent biopiracy, have no legitimacy under DR-CAFTA. As GRAIN (2006) points out, this is not surprising as the United States is not a signatory to the CBD; because disclosure of origin, consent and benefits are not demanded, patent applicants have carte blanche to engage in bioprospecting as they choose.

Biosafety regulations affecting seeds

Honduras is a signatory to the Cartagena Protocol on Biosafety (to the CBD) and a National Committee on Biosafety exists. The latter is made up of representatives of the National Autonomous University of Honduras, the Panamerican Agricultural School (Zamorano), public health, the Ministry of Environment and Natural Resources, the National Service of Plant and Animal Health (SENASA), the Directorate of Science and Agricultural/Livestock Technology, the focal point of the Codex Alimentarius in the Ministry of Agriculture and Livestock, the Honduran Council of Science and Technology and the Standard Fruit Company.

The first request to use biotechnology in Honduras came from the Standard Fruit Company in 1996. In 1997, Monsanto began testing Bt-maize. As there were no regulations governing biotechnology at the time, the Seed Certification Department of the Ministry of Agriculture and Livestock introduced the Biosecurity Regulation with Emphasis on Transgenic Plants. This was approved in 1998.

However, the legal basis for the regulation is the Phytosanitary Law of 1994, which excluded mention of transgenic plants. The 1994 law is rooted in phytosanitary seed protection dating back to the Seed Law of 1966, subsequently modified in 1980. A further modification of the Phytosanitary Law (Decree 344-2005), in accordance with DR-CAFTA, published in 2006, makes no reference to transgenic plants. According to interviews conducted by Galeano (unpublished) with personnel at the SENASA, the agency in charge of biotechnology regulation, the SENASA has no equipment for analyzing DNA and, therefore, cannot handle biosafety concerns posed by the introduction of transgenic plants and seeds into the country.

In the absence of a law, decisions are made on an *ad hoc* basis by the Committee on Biosafety with the support of the Organismo Internacional Regional de Sanidad Agropecuaria, together with the Centro Agronómico Tropical de Investigación y Enseñanza (CATIE) and the Inter-American Institute for Cooperation on Agriculture (IICA). However, the Honduran Ministry of Environment and Natural Resources, which is in charge of applying the Cartagena Protocol on Biosafety, represents the country at international meetings. In 2006, interviews with SENASA personnel on the use of transgenic seed in Honduras were met with flat denial, even though it was already widely known that the use of Bt-maize was well advanced in several departments (Galeano 2006). At the time, SENASA personnel were prohibited from expressing any opinion on biotechnology because of what was perceived to be the sensitive nature of the topic.

Three years later, 14,755 hectares of transgenic maize—Bt, RR and Herculex 1—were being planted by commercial growers in departments across the country and Honduras had become the first country in the region to embrace genetically modified crops, which it openly regarded as a solution to the food crisis (Charles 2008). It was also producing genetically modified maize seed for export to

Argentina, Colombia and the United States. Thus Honduras has moved from covert production to being a regional leader and exporter of seed, despite the absence of a national biosafety law.

Since 2005, the Pan American Agricultural School (Escuela Agricola Panamericana - Zamorano (EAP-Zamorano)), has been conducting biotechnology research producing about 80 hectares of parental Bt and RR maize seed per year. It is contracted by various seed companies to undertake biotechnology maize research on their behalf and to monitor the seed with regard to local growing conditions and pathogens. Monsanto has been selling Bt-maize seed in Honduras since 2003 and Pioneer entered the market in 2008 with its Herculex 1 seed.

EAP-Zamorano is also the primary producer of bean seed. The bean research program provides breeder, foundation and registered seed, while certified bean seed is sold through its private company, Zamoempresa de Cultivos Extensivos. Hondugenet, a seed company that changed from parastatal to private status in 1992, sells certified and commercial bean seed (Pejuán 2005). Currently, it is the only independent national seed company in existence, where there were 14 commercial seed suppliers in the recent past (Galeano unpublished interviews).

The Directorate of Science and Agricultural/Livestock Technology sells only very small amounts of foundation, registered and certified seed. Both Hondugenet and Zamoempresa de Cultivos Extensivos contract with local farmers to produce their seed, and phytosanitary regulations are enforced by the biosafety agency, SENASA (Pejuán 2005: 56–60). However, according to Hondugenet officials, SENASA rarely regulates seeds produced by EAP-Zamorano, while Hondugenet (as well as past seed suppliers) faces stiff certification requirements (Galeano unpublished interviews). Thus, Hondugenet argues, it is forced to operate in an unfair and overly regulated marketplace which is effectively handing EAP-Zamorano a monopoly in the bean seed market. Moreover, as a teaching institution, EAP-Zamorano is tax exempt and, therefore, operates in the private seed market at a considerable advantage.

In short, in recent years multinational seed companies have come to dominate the Honduran maize market. The registered, certified bean market is controlled by EAP-Zamorano, while Hondugenet fights an uphill battle to hold on to its share of the market for commercial bean seed.

The implications of international agreements for farmers' seed supply

Regional and local NGOs have expressed concern about the biosafety implications of genetically modified plants and their effect on biodiversity. Certainly, the risk of hybridization between genetically modified maize and farmers' open-pollinated varieties is very real. However, maize production has also lost its appeal for larger farmers because of the market disincentives discussed above. The potential for higher yields through biotechnology is viewed by commercial growers as the only way to be competitive, especially as trade barriers to yellow maize are reduced.

Small farmers will continue to produce maize to cover subsistence needs and, depending on options available to them, for sale on the market.

At present, genetically modified maize is unprotected by patent because of government reluctance to legislate variety protection. Thus it is beginning to be used extensively by small lowland and valley farmers who are sharing it widely with family and friends. Those who fail to obtain it through social networks often do so through stealth. Enforcement of intellectual property rights, even if such rights existed, is unlikely to be effective in this environment.

Lowland agrobiodiversity has likely already been eroded by the dominance of improved commercial plant varieties. In a study of 20 low-altitude communities on the Atlantic coast in 2009, only six maize and six bean landraces were found. This contrasts with the much broader use and wider variety of landraces in evidence in most upland communities, where improved varieties have failed to make inroads. Genetically modified varieties are unlikely to take hold in these areas as they have not been bred for upland conditions where most of the smallest farms are located. Thus, although contamination of farmers' landraces and locally improved seed by genetically modified maize cannot be ruled out, the risk of displacing small farmers' seed varieties seems less of a concern. If this prediction is correct, agrobiodiversity will not be much affected.

A more serious concern for small farmers seems to be the risk of biopiracy. Specific traits in the landraces that farmers have conserved or improved through participatory plant breeding (PPB) might become transferable to materials protected under UPOV-91. Should the Law for the Protection of Plant Varieties eventually pass, small farmers are unlikely to benefit from protection of farmers' rights. As the history of local battles against mining companies and other extractive industries in Honduras shows, corporate interests almost always prevail over those of local communities. Corruption—in which Honduras ranks 107 of 158 countries along with Zimbabwe, Palestine and others (Transparency International 2005)—makes it easy for those with the financial means to win almost any judicial case, however unjust the outcome may be. Those with resources simply "buy off" the necessary authorities without ever proceeding to court. In Honduras, disputes between the rich and poor mean that the poor are almost always losers.

Participatory plant breeding in the new globalization context

Relations between plant breeders and farmers are likely to be unequal in all societies. Given the socioeconomic inequalities and the monopoly exerted by EAP-Zamorano over access to genetic materials for research, in Honduras, the disparity is greater than elsewhere. Thus benefit sharing among those involved in PPB will require willingness on the part of Zamorano's scientists, as it is not something that can be enforced through legal or extra-legal means. That said, PPB is unlikely to be undertaken by breeders unless they are open to recognizing farmers' contributions. It is most likely to be supported by breeders who view it as beneficial to the breeding process and, therefore, as supportive of their reputation as breeders. Thus, although some recognition of benefit sharing is part and parcel of PPB, how far this recognition goes is certainly something that is up for debate.

DR-CAFTA is already having an impact on the seed market: foreign seed suppliers now dominate the maize market and may soon make inroads into the bean market as well. If transnational corporations succeed in penetrating the Central American bean market, national suppliers of bean seeds, such as Zamorano and Hondugenet, will almost certainly see their share of the market decline. Zamorano, which is a research institution supported by public funds as well as a private commercial company, may find itself ever more in a conflict of interest, positioned as it is between production of public goods on one hand and an increasingly competitive market for private goods on the other. In particular, "genetic leakage" from research with upland farmers engaged in PPB, into research for commercial gain, could potentially pose a problem and will require careful monitoring to prevent it.

Benefit sharing: the experience of the comités de investigación agrícola local in Yorito

The first experience with PPB in Honduras was carried out in conjunction with a local NGO, Fundación para la Investigación Participativa con Agricultores de Honduras (FIPAH) and farmer research committees, comités de investigación agrícola local (CIALs). FIPAH originated from a pilot project set up by the International Center for Tropical Agriculture (CIAT) in 1993 and later supported by IDRC through the University of Guelph. Since 2000, it has been funded by a Canadian NGO, USC-Canada. Additional funding has been provided through the development fund of Norway's mesoamerican PPB program and through a small grant from CGIAR's system-wide Program on Participatory Research and Gender Analysis.

PPB was first undertaken by CIALs in the municipality of Yorito in northeastern Honduras in 1999–2000. Yorito is one of the poorest municipalities in the region and its inhabitants are engaged almost entirely in subsistence agriculture. In the valleys, some irrigated agriculture and cattle production is possible, but, on the hillsides, farming families lead a precarious existence. Most suffer from acute food insecurity during *los junios*, the "hungry season" that generally begins in June when stocks from the previous harvest have run out and continues through to August when the next harvest begins.

After four years (1996–99) of testing breeder-improved varieties of maize and beans, CIAL researchers had not achieved any major increases in yield at upland locations. Although breeder-improved materials generally outperformed local materials at lower altitudes, at upland locations where the poorest farmers lived, they failed to produce a significant increase in yield. Indeed, in farmer-led experiments at locations above 1,000 m, local varieties produced the best outcomes, despite low yields and susceptibility to certain diseases (Humphries *et al.* 2005). These findings, coupled with the availability of two small grants to

Zamorano for PPB, led to the development of bean and maize improvement programs in Yorito. Since FIPAH began its program in 1999, 16 varieties have been improved at its field locations in various regions of the country.

The experience with CIALs in the municipalities of Yorito and Sulaco provided examples of various degrees of farmer and breeder involvement in PPB. These experiences can be placed on a continuum, listed below with examples:

- **Farmer-led**—Through intensive selection, farmers improve a local landrace and release it with a new name. This process is supported by a local NGO.
- **Farmer–breeder collaboration**—At the request of farmers (and in conjunction with donor support for PPB), the breeder crosses a local landrace with improved materials and returns segregating populations to farmers for intensive selection. Farmers select certain lines and release varieties with new names. The NGO acts as an intermediary between the breeder and farmers, providing detailed feedback on the trials to the breeder and technical assistance to farmers.
- **Breeder-led**—The breeder provides farmers with improved lines, e.g. through regional trials, and farmers select certain materials that are not selected elsewhere in the country or region. These become local varieties and are given local names. The NGO provides technical support to the farmers and information for use in analyzing data from regional trials to the breeder.

Farmer-led PPB

CIAL staff in the community of Santa Cruz decided to improve their local maize landrace, Capulin, with FIPAH support. The local variety of white maize grows well between 1,000 m and 1,800 m, but suffers from problems of synchronization of silking and pollen shedding leading to low yields. Lodging is also a problem, as tall plants are favored by post-harvest selection of the biggest cobs without regard for plant architecture. Thus, a selection program was designed to reduce the height of the maize, in particular to lower the location of the cobs on the stalk, and to improve synchronicity.

PPB involved a mass composite approach, requiring collection and sowing of selected seed from four communities. Farmer researchers from Santa Cruz led the process, which involved participatory field and post-harvest evaluation of seed over four cycles of planting in a pattern of recombination plots. Half the seed from each selected cob was retained for planting only after the results of the previous test cycle became known. In this way, seed selected in the test cycle remained uncontaminated by pollen from less-desirable plants discarded during the test cycle. In the recombination plots, seed from the retained half cobs (designated as "female") were planted between rows of "male" maize. The latter comprised a mixture of the retained, selected seed and served to produce pollen to fertilize the rows of "female" plants. This process was repeated twice, in 2001 and again in 2002, in each case using a test cycle that relied on irrigation, allowing the recombination plots to be sown during the traditional growing season. In 2003, the best plants were sown for seed multiplication.

The selection process led to a height reduction of maize plants from 3.1 m to 2.7 m. Yields increased by as much as a tonne per hectare and average yield increased by about 25%. These results were partly a consequence of the shorter stature, which allowed farmers to increase the planting density, as well as greater yield per plant. The chosen variety, Capulin Mejorado, was released at the municipal level in October 2005.

This research was carried out by the farmers with support from FIPAH; Zamorano played no obvious role in the process. Using small grants, a young agronomist was hired to support the PPB process in Yorito. Although he was employed by Zamorano, he became a member of FIPAH and was regarded as such by everyone. Lack of direction at the outset led to some loss of time as the team searched for the most appropriate method. In April and again in August 2006, members of the Yorito seed committee discussed benefit sharing between the three parties—Zamorano, FIPAH and farmers—and came up with a hypothetical breakdown: farmers would get 70%, FIPAH 30% and Zamorano 0%. The decision to include FIPAH as a beneficiary related to the shared experience of working together. How the breakdown was calculated was not made clear. The principal author left the room, as requested, so that CIAL members could freely discuss the allocation of benefits. As in the examples below, Zamorano was not contacted to discuss this breakdown and farmers were reluctant to do so.

Farmer–breeder collaboration in PPB

Seed from the local landrace, Capulin, was collected from four upland communities and sent to Zamorano for crossing with two improved maize varieties (DICTA Guayape and HB104). Zamorano returned second-generation material to CIAL Santa Cruz, whose members selected materials over the next three generations.

Zamorano's role was selection and intra-population recombination activities at its experiment station. The cross with Guayape was not successful and was subsequently discarded. However, hybridization with HB104, a short-stature variety frequently used at lower altitudes, produced a new variety, Santa Cruz, which was well accepted. Through the PPB process, plant height was reduced from 3.1 to 2.6 m and cob height from 1.8 to 1.3 m; yields increased by around half a tonne per hectare over the unimproved landrace parent. The new Santa Cruz variety is adapted to conditions from 900 to 1,200 m above sea level, considerably above the normal range of HB104. It was released at the municipal level in October 2005.

Members of the seed committee designated the following allocation of benefits: farmers 50%, FIPAH 30% and Zamorano 20%. This recognized the work of Zamorano in providing the elite genes involved in hybridization. Nevertheless, the farmers put more weight on the value of local germplasm and their own work, as well as FIPAH's support of that work.

In the case of the bean variety, Mazucalito, after four years of rather unsuccessful experimentation with breeder-improved varieties, upland CIALs turned to improving their own most popular landrace, Concha Rosada, through PPB. This local variety is a small red bean which was probably introduced in the 1980s, then adapted by local farmers to upland conditions. However, despite its adaptation, it is affected by a number of diseases leading to low yields. In addition, it has several other features, such as trailing architecture and uneven ripening, that farmers dislike and hoped to eliminate. Seeds were collected from neighboring communities, selected by the CIALs and given to Zamorano for crossing with elite lines. Five breeder materials were employed (Tio-Canela-75, SRC 1-12-1, MD 23-24, SRC 1-1-18 and UPR 9609-2-2).

CIAL members (30 men and 23 women) from four communities (Mina Honda, Santa Cruz, La Patastera and Chaguitio) initially received two populations from the cross of Concha Rosada with Zamorano materials (the maternal parent; subsequent populations in which Concha Rosada acted as the paternal parent did not produce beans of a desirable color and testing was abandoned). These comprised 120 families in the third (F3) generation.

Initially, all the materials were kept at a collective site in Mina Honda, 1,350 m above sea level. However, farmers from the other three communities elected to take their selections back to their own communities to run local trials involving replicate plots. The best plants from the local community trials were subsequently used in comparative F6 trials conducted in replicate plots across participating communities.

Zamorano, which ran on-station trials concurrent with those of the farmers, also contributed its best varieties to the F6 Yorito trials. Farmers eventually selected four materials for subsequent testing in adaptive trials (F8), none of which originated from the Zamorano trials. One of the four lines, named Macuzalito by CIAL members, was selected from the F8 trials and released at the municipal level in August 2004. In 2009, a study in 106 nearby communities showed that Macuzalito was being used in 48 communities by 191 farmers, the second most widely used variety in the surveyed communities (ASOCIAL-FIPAH 2010).

Throughout the PPB process, the farmers received extensive agronomic support from FIPAH. None of them had carried out such work before, and they had to learn how to select for characteristics that might vary from one generation to the next. This was also a new process for FIPAH and a good deal of mutual learning took place. In total, FIPAH provided 97 training sessions to the CIALs involved in the bean breeding process. The training took place at upland locations and required a good deal of travel time. FIPAH provided Zamorano with detailed information on the individual community trials, but Zamorano's role was largely invisible at the local level.

Farmers decided on the following breakdown of benefits: farmers 50%, FIPAH 30% and Zamorano 20%. Thus, benefit sharing followed the same breakdown as in the previous case, and the procedure followed by members of the committee was the same. Subsequent to the release of Macuzalito, Zamorano further improved the seed.

Breeder-led PPB

Zamorano has consistently supplied CIALs with new materials for testing as part of national and regional trials. These have sometimes been supplied early in the testing process (F4–F6 stages) or for adaptation once materials have stabilized (after F6). In return, FIPAH provides feedback to Zamorano on the results of the trials.

A number of locally released varieties have been developed through this testing process. One notable case is the bean, Cedron. In 1999, a group of farmers, who were formally organized as CIAL Chaguitio the following year, was provided with a set of 15 F6 materials by Zamorano as part of a regional trial. During three rounds of selection (F6–F8), the CIAL eventually chose one line, Cedron. This material had never been released nationally by Zamorano; however, in 2007, the CIAL released it at the municipal level. A 2009 study in 106 communities showed that it was being used by 201 farmers in 13 communities and was the most widely used bean variety in the surveyed communities (ASOCIAL-FIPAH 2010).

In this case, the hypothetical distribution of benefits was farmers 50%, FIPAH 20% and Zamorano 30%. Farmers decided to allocate more benefits to Zamorano in this case than in the previous two. When asked if they still felt that they deserved 50% of the benefits and whether Zamorano should not be afforded a larger proportion as it had provided all the germplasm, CIAL members responded that they felt it to be the correct breakdown. They reasoned that if they had not taken the trouble to work with the materials, Zamorano would not have used them and they would have been wasted. In other words, they were not prepared to place a great deal of extra value on Zamorano's role over that accorded it in the previous examples in which their own germplasm was used alongside Zamorano's. Rather they felt that their labor was still the vital ingredient. FIPAH was accorded a somewhat smaller percentage of the benefits in this case, although their support was similar to that in the examples described above.

Benefit sharing and intellectual property rights

It is clear from farmers' allocation of the hypothetical benefits from PPB that they are not prepared to give breeders a significant portion, even when they provide the plant materials: the breeder, who is generally out of sight, is largely also out of mind. Instead, farmers regard their labor and skills as the main components of PPB. Indeed, human resources appear to be more important to them than rights over local germplasm. Thus, the longer and more complex the process of PPB, the more farmers are likely to feel they have rights over benefits ensuing from it, independent of where the germplasm originated.

From the breeders' perspective, however, farmer selection of breeder-provided plant material is more likely to be viewed as validation of their skills as breeders than as farmer creativity, especially if the breeders have little opportunity to witness the skill and effort farmers put into the process. Just as farmers remain unaware of the resources (both human and financial) invested in plant breeding, breeders who rarely stray from the research station are likely to be similarly uninformed. As Zamorano has only one breeder, who also teaches at the school, it is clearly difficult to maintain a significant presence in the countryside.

Farmers and breeders need organizations that can straddle the divide between them and help mediate discussions around benefit sharing. NGOs that have their feet firmly planted in farmers' fields and also understand the complexities of formal-sector plant breeding are perhaps best positioned to provide a credible perspective on the appropriate sharing of potential benefits. An understanding of ABS with regard to genetic resources should be arrived at before PPB is undertaken to prevent disagreements over ownership later on.

Beyond seeking agreement between the parties most closely involved in PPB, there is also a need to establish conditions affecting farmers' rights more broadly. As mentioned, Honduran CIALs involved in PPB have sought to maintain rights over their varieties through municipal release of the materials. This has involved a ceremony, attended by the mayor or deputy mayor and other municipal officials, farmers, breeders, NGOs, etc., during which a special municipal act is read stating that the seed has been created by CIAL members for use by high-altitude farmers and is not intended for commercial purposes without prior consent. In each case, the seed is blessed by the local priest, then ceremonially presented to the municipality.

The support this process provides for farmers' rights is more symbolic than substantive. An exchange with another NGO is illustrative. Following the release of Macuzalito in 2004, a handful of CIAL members sold seed to a large international NGO, which had a local office close by. The NGO, in turn, parceled the seed out to farmers with irrigated land in the valley for the purpose of seed multiplication. These farmers were paid to produce seed, which was ultimately used to support the NGO's clients in the region.

After hearing about the transaction, we wrote to the NGO's country director several times asking whether they would consider donating a small percentage of the large-scale farmers' payment for the seed to the local association of CIALs in compensation for five years of work. Our concern was that unless some payback was received for their work, CIAL members would lose interest in PPB over time. The NGO refused to respond to the request.

In this case, the use of the seed by the NGO does not violate the terms of the municipal act as the seed was intended to support poor farmers. Nevertheless, the large-scale farmers received financial benefit at the expense of five years of work by an impoverished group of farmers. Hillside farmers with tiny properties are not, for the most part, able to engage in large-scale seed production. Moreover, seed production must be accomplished during the dry season, so that fresh seed is available for planting when the rains come. This requires irrigation, something that most poor farmers do not have. Had the encounter been with a corporation intent on large-scale commercial production, the farmers' sense of impotence might have been more acute.

Artisan seed production

Since they began the PPB process in 1999, CIALs, supported by FIPAH, have produced 16 varieties. Many of these have been released at the municipal level, as described above; others are still undergoing testing at various locations. As a result of their PPB involvement, some CIAL members have begun to produce seed as a means to earn additional income. The biggest constraint facing them, however, is local farmers' traditional unwillingness to pay a premium for this seed.

In 2006, before the recent global food price surges, beans were sold locally at 2 lempiras/pound (about US$0.10). CIAL members, who produced clean bean seed involving a considerable amount of labor, were struggling to get 5 lempiras/pound. Meanwhile, Zamorano was charging 16 lempiras/pound for its seed, which was mainly being purchased by commercial bean growers. Small seed producers, intent on multiplying PPB seed so as to distribute it to neighboring communities, are faced with the reluctance of the poorest farmers to spend scarce resources on improved seed, as opposed to simply using seeds from their own crops.

Selling seed to the NGO community, as in the example above, provides a solution. In this case, the NGO subsidized the price of seed, underwriting the seed producers' costs. As most CIAL members are more intent on getting their plant material to other poor upland farmers than making a profit, expectations surrounding income generation are quite modest; most simply share their seed with neighbors and friends. However, whether seed is produced for income generation or for sharing, CIAL members emphatically do not want to see their seed being sold by other organizations at commercial prices without any benefits flowing back to them.

Thus, if Zamorano were to begin to sell CIAL-generated PPB seed, whether or not it contained genetic resources originally supplied by Zamorano, it would be considered a serious breach of the unwritten agreement between Zamorano and farmer collaborators, as the hypothetical allocation of benefits by farmers in the previous discussion makes clear. To avoid future misunderstandings, a discussion of ABS from genetic resources needs to be undertaken by the various participants in the PPB process.

Conclusions

The nature of agricultural production in Honduras, as well as in the rest of Central America and the Dominican Republic, is being changed by DR-CAFTA, although it is still too soon to document how change is occurring. Examples from other countries in the aftermath of free trade agreements, however, point to a loss of agricultural livelihoods associated with cheap grain imports and downward pressure on local prices. Small Honduran farmers are not exempt from the agreement in spite of certain provisions designed to protect them. In a country where half the population is still mainly engaged in agriculture, the impact of DR-CAFTA will be profound.

Farmers' rights are unlikely to be defensible, notwithstanding the ITPGRFA. Social inequality and corruption make it almost impossible for the poor to be fairly represented under the law. Under DR-CAFTA, farmers' indigenous seeds and knowledge are considered patentable commodities and prior informed consent, disclosure of origin and benefit sharing are considered unnecessary before applying for a patent.

Although the government has delayed passing laws protecting intellectual property rights to plant material, pressure from multinational corporations is likely to intensify, particularly in light of the amount of unregulated sharing of genetically modified maize seed that is already occurring. If plant varieties become patentable, germplasm conserved by farmers or created by them jointly with other parties will be difficult to protect from practices (such as "bioprospecting") that permit others to benefit unfairly at their expense.

Zamorano has a good deal of authority within the region and offers the best source of protection against theft of farmers' materials. However, Zamorano also runs a commercial seed operation. Thus it is necessary for farmers and NGOs to discuss benefit sharing openly with Zamorano to avoid future friction over ownership rights. We hope that by providing a panorama of the changing nature of the seed sector both nationally and regionally, and by locating farmer-breeders within this context in this chapter, we have taken a step toward giving farmers a stronger footing in these discussions.

References

ASOCIAL-FIPAH (Association of CIALs—Fundación para la Investigación Participativa con Agricultores de Honduras) (2010) Estudio de aceptabilidad de la variedad de frijol Macuzalito, Yorito. Unpublished document.

Centro de Desarrollo Humano (2005) Investigación sobre los efectos de CAFTA-RD en el sector rural de Honduras. CDH, Tegucigalpa, Honduras.

Charles, Dan (2008) Honduras embraces genetically modified crops. NPR, Washington, DC. Available at: www.npr.org/templates/story/story.php?storyId=93310225 (accessed 21 February 2011).

Cooper, D. (2003) International treaties relevant to the management of plant genetic resources. In *Conservation and Sustainable Use of Agricultural Biodiversity*. Vol. 3 Ensuring an Enabling Environment for Agricultural Biodiversity. Centro Internacional de la Papa—Users' Perspectives With Agricultural Research and Development, Los Baños, Philippines. pp. 461–74. Available at: www.eseap.cipotato.org/upward/Publications/Agrobiodiversity/pages%20461-474%20(Paper%2057).pdf (accessed 15 February 2011).

Galleano, R. (n.d.) Consultaría sobre las implicaciones de los acuerdos internacionales que afectan la regulación de los sistemas de semillas en Honduras. Unpublished interviews.

Gomez, A. (2009) Honduras: agricultural biotechnology annual (Global Agricultural Information Network report HO9007). United States Department of Agriculture, Washington, DC, USA. Available at: http://gain.fas.usda.gov/Recent%20 GAIN%20Publications/AGRICULTURAL%20BIOTECHNOLOGY%20ANNUAL_ Tegucigalpa_Honduras_7-20-2009.pdf (accessed 21 February 2011).

GRAIN (Genetic Resources Action International) (2003) Right to patent vs. right to freely use: TRIPS, UPOV and farmers' rights. In *Conservation and Sustainable Use of Agricultural Biodiversity*. Vol. 3 Ensuring an Enabling Environment for Agricultural Biodiversity. Centro Internacional de la Papa—Users' Perspectives With Agricultural Research and Development, Los Baños, Philippines. pp. 514–21. Available at: www.eseap.cipotato.org/upward/Publications/Agrobiodiversity/pages%20514-521%20(Paper%2061).pdf (accessed 21 February 2011).

GRAIN and Rodriguez Cervantes, S. (2006) FTAs: trading away traditional knowledge. GRAIN, Barcelona, Spain. Available at: www.grain.org/briefings/?id=196 (accessed 21 February 2011).

Humphries, S., Gallardo, O., Jimenez, J. and Sierra, F. (2005) Linking small farmers to the formal research sector: lessons from a participatory bean breeding programme in Honduras (Agriculture Research and Extension Network paper 142). Overseas Development Institute, London, UK. Available at: www.odi.org.uk/AgREN/papers/agrenpaper_142.pdf (accessed 21 February 2011).

Jansen, H.G.P., Siegal, P.B., Alwang, J. and Pichon, F. (2005) Geography, livelihoods and well-being in rural Honduras: an empirical analysis using an asset-base approach. Presented at the Conference on Poverty, Inequality and Policy in Latin America, University of Goettingen, Germany, 14–16 July.

Pejuán, W. (2005) Constraints to and opportunities for developing a bean seed production and marketing system in Honduras. MSc thesis, Michigan State University, East Lansing, MI, USA.

Transparency International (2005) Corruption perceptions index 2005. Transparency International, Berlin, Germany. Available at: www.transparency.org/policy_research/surveys_indices/cpi/2005 (accessed 21 February 2011).

World Bank (2004) Honduras: drivers of sustainable rural growth and poverty reduction in Central America (report 31192-HN). World Bank, Washington, DC, USA.

Zappacosta, M. (2005) Honduras: market profile for emergency food assessments. World Food Programme, Emergency Needs Assessment Branch, Rome, Italy. Available at: http://documents.wfp.org/stellent/groups/public/documents/ena/wfp086535.pdf (accessed 21 February 2011).

9 China

Designing policies and laws to ensure fair access and benefit sharing of genetic resources and participatory plant breeding products

Yiching Song, Jingsong Li and Ronnie Vernooy, with the collaboration of the Guangxi-based research team of plant breeders and farmers and the Beijing-based policymakers

Challenges facing the new China

As the largest developing country in the world, with its very diverse culture, economy and ecology, China is experiencing a dramatic transition period in many spheres. China's entry into the World Trade Organization (WTO) marks an important event in this period and has accelerated the economic transition process. The opening of China's economy to foreign enterprises and its integration with the global economy is having profound but different socioeconomic impacts on different groups of people in different sectors and different areas.

The impact of these changes on vulnerable groups, such as smallholder farmers and their communities, and the implications for policymaking and the reform of regulations and laws has been drawing increasing attention from policymakers as well as researchers. The most important issue is unbalanced development, in terms of gaps between urban and rural regions, industry and agriculture, the east coast and remote western areas, economic development and protection of the environment. This is leading to challenges, such as extreme rural poverty and inequality, feminization and aging of agriculture, severe environmental degra-dation and erosion of biodiversity (Song and Vernooy 2010a). Ensuring China's national food security has recently been added to these concerns (for an "early" warning, see Huang 2003).

Extreme rural poverty

With the adoption of a broad program of rural economic reforms beginning in 1978, the Chinese rural economy has experienced rapid growth, and China has been widely recognized for its achievements in reducing extreme poverty since then. Nevertheless, about 30 million people still live below the extreme poverty line (i.e. income below US$0.5 a day (State Council Office of Poverty Alleviation 2007)), and they make up the majority of the food-insecure population. Based on the global standard of poverty, i.e. US$1.25 a day, the poor population numbers over 100 million (Juan Zhang 2010). They are mainly farmers in remote upland areas of southwest and northwest China that are agro-ecologically diverse, resource poor and risk prone. They are mainly smallholder farmers with an average area of less than 0.2 ha. Although these poor have land use rights, in most cases the land is of such low quality (with many rocky areas), it is not possible to achieve subsistence levels of production. Consequently, most poor people consume grain and other subsistence foods beyond their own production means and have been negatively affected by the increased price of these products since the reforms. Ethnic minorities represent a highly disproportionate number of the rural poor.

Feminization and aging of agriculture

Recent studies in China have revealed an overall increase in out-migration from rural to urban areas, especially from poorer areas. These studies also show that far fewer women are moving, resulting in increasing "feminization" of agriculture in the last decade, especially in the poor western areas (Zuo and Song 2002, UNDP 2003, Song and Zhang 2004). More and more women are being left alone at home, and they have become more and more engaged in agriculture and natural resource management. Women constitute about 70–80% of the agricultural labor force in most provinces, especially in the west and southwest (Song, Zhang *et al.* 2006, Song and Vernooy 2010b). They are mainly middle-aged women with limited education.

Environmental degradation and loss of biodiversity and related traditional knowledge

Southwest and northwest China are rich in terms of culture and diversity. However, in the last few decades, the environment and natural resources in these regions are suffering rapid degradation as result of overexploitation and inappropriate interventions. Smallholder farmers and their farming communities are finding it more and more difficult to conserve and enhance agricultural biodiversity. Their traditional knowledge and local practices of experimentation and innovation are under stress. Biological diversity, especially the landraces in farmers' fields, are disappearing at an accelerating speed (Zhang *et al.* 2010). This trend is threatening the livelihood and security of the poor and national agricultural sustainability and food security in the long run.

In the last ten years, crop landraces in the southwest have decreased rapidly. For example, a recent survey in Guangxi, Yunnan and Guizhou has revealed that although about 90% of respondent households were cultivating maize landraces in 1998, this proportion decreased to 73% in 2003 and 56% in 2008. Farmers are turning to hybrids and buying seeds from markets (CCAP 2009). This is mainly the result of market forces backed by government policies and interventions, leaving less and less space and options for farmers and farmers' own seeds systems.

Institutional and legal issues and constraints on seed systems and public services

Maize research in China is well organized and has produced good results, but it has been carried out mainly under favorable growing conditions. Less arable regions, including Guangxi, Yunnan and Guizhou, have not been served well, partly because of the prevailing assumptions of maize breeders: the belief that farmers are less knowledgeable than breeders, that selection must be done under optimum conditions, that cultivars must be genetically uniform and widely adaptable over large geographic areas and that landraces and open-pollinated varieties (OPVs, such as those still found in the southwest) must be replaced by high-yielding varieties to ensure national food security. Biodiversity, farmers' diverse livelihoods and their contribution to crop improvement have been largely ignored (Zhang *et al.* 2010). Most hybrid varieties are unable to adapt to the conditions in remote mountainous areas, as in Guangxi and other southwestern provinces. They are also susceptible to diseases, pests and drought (Figure 9.1).

Figure 9.1 Farmer-improved landraces (back) survived the severe spring drought of 2010, but hybrid maize varieties growing on the land in Guangxi did not (front)

Source: Photo by Cheng Weidong

In marginalized areas, farmers' seed systems continue to play a major role in the seed supply, while maintaining the diversity that is essential to sustain the livelihoods of all farmers (and the country at large). An impact study on the dissemination of new maize varieties in Guangxi (Song 1998) revealed that, in the southwest remote mountainous area, more than 80% of the seed supply was from farmers' own seed systems. A more recent study (CCAP 2008) showed that 30–40% of the maize-growing area in the three southwestern provinces is planted in OPVs, which rely on the seeds farmers save. OPVs also still cover more than 70% of the mountainous parts of those areas. However, there is little or no recognition of farmers' contribution to policies and laws related to agricultural development, let alone concrete support.

The agricultural extension system is an example of this lack of support. During the country's rapid transformation from a planned economy to one that is more market-oriented, the extension system became paralyzed and obsolete. In the 1990s, the whole system almost collapsed: no real service delivery took place, few or no innovations reached farmers, connections with other rural development agencies were ineffective or nonexistent, the capacity of the extension system had not been maintained and most staff dedicated time and energy to tasks other than serving farmers and contributing to sustainable rural development (Zhang *et al.* 2010). Many local extension stations became seed, fertilizer and pesticide shops, and farmers often did not even realize that they were government-run enterprises. Concern over farmers' own seed systems was mostly absent.

To address these challenges, in 1999, a novel action research initiative was started by several groups of women farmers, a number of rural villages and township extension stations, two formal plant breeding organizations in the Chinese national agricultural research system, and the Center for Chinese Agricultural Policy (CCAP). The CCAP has provided coordination and guidance with regard to research design, implementation and use of the research results. In recent years, other villages have joined. In 2008, similar work started in the neighboring provinces of Yunnan and Guizhou led by CCAP, the Institute of Crop Science (Chinese Academy of Agricultural Sciences) and the Ministry of Agriculture. What started as a small initiative focusing on improving maize varieties in Guangxi has grown into an effort to revitalize rural development by addressing not only crop production, but also sustainable agriculture, including ecological (e.g. switching to organic agriculture), sociocultural (e.g. supporting farmers' own cultural initiatives), economic (e.g. improving marketing options) and political aspects (e.g. linking farmers to decision-makers in science, higher education and policy development). For a detailed account, see Song and Vernooy (2010a).

This team aimed to change things through a sustained, action-oriented, participatory research effort. The experience of a decade illustrates the successes and challenges of linking community-based action research with policy- and law-making processes by increasing efforts to engage key decision-makers in the rural development policy arena at local, provincial and national levels (Vernooy and Song 2010). We discuss these challenges in more detail below, but first provide a short overview of the participatory plant breeding (PPB) and biodiversity conservation activities that have been at the heart of the work. More details about the technical aspects of PPB in Guangxi can be found in Song *et al.* (2010).

Pioneering PPB

The main aim of the PPB initiative is to establish cooperative and complementary relations between the formal seed system and farmers' systems. Cooperation is necessary to provide opportunities for the empowerment of farmers—mainly women farmers, as most men have migrated to the cities. The farmers become active partners in plant breeding, on-farm biodiversity management and seed marketing (Ashby 2009, Song and Vernooy 2010b).

The PPB method is adapted to the local context. The work of the entire team, including the farmers, builds on local women farmers' maize breeding experience and expertise developed over many years (Song 1998). At the same time, the team involves and seeks knowledge and expertise from formally trained plant breeders. Crop improvements are made through a number of crossing techniques and various variety selection processes, which involve detasseling, mass selection and line selection by farmers with support from breeders. Breeders use more complex methods in the fields of the Guangxi Maize Research Institute (GMRI) in Nanning. The work has covered a range of parallel activities over a number of years using various methods to identify parental materials (through participatory variety selection), improve populations (involving local and formal-system genetic materials) and select further to obtain individual varieties. Trials in six villages and at the GMRI include both PPB and participatory variety selection. These trials are evaluated by both breeders and farmers after each cycle and, subsequently, new designs are discussed and agreed jointly. The trials allow for comparisons in terms of locality, approach, objectives and the types of varieties tested (Song 2003, Song and Jiggins 2003, Song, Zhang S. *et al.* 2006, Song *et al.* 2010).

New maize varieties

As a result of a series of discussions among farmers and formal plant breeders, jointly and separately, the field experiments have targeted four types of OPVs and landraces, i.e. "exotic" populations (from abroad), farmers' "creolized" varieties (developed by breeders but further adapted by farmers, sometimes by crossing them with landraces), farmer-maintained landraces and formally conserved landraces. So far, more than 80 varieties have been used in trials at the GMRI station and in the villages. Based on ten years of experimentation, four farmer-preferred varieties have been selected and released in the research villages. They have also spread beyond these villages. In addition, five varieties from the International Maize and Wheat Improvement Center that were showing increasingly poor results have been adapted locally. Another five landraces from the trial villages have been improved thanks to the joint efforts of farmers and formal breeders. Agronomic traits, yields and palatability of all these varieties are satisfactory and they are better adapted to the local environment (CCAP 2004, Song, Zhang S. *et al.* 2006).

A women-farmer-improved variety, known locally as New Mexico 1 (i.e. Xin Mo 1) has been tested over a number of cycles and certified by the formal breeding institution. Its robustness and taste make it very popular, and it is now widely used locally. Farmers from neighboring areas, who have heard about this variety, are coming to learn more and to ask for seeds. In the research area, varietal diversity is increasing. Meanwhile, formal breeders have identified in farmers' fields a number of useful breeding materials with a valuable, broad genetic base.

After several years of PPB, the team has isolated five varieties. They are:

- **Xin Mo 1** (New Mexico 1), an OPV, was derived from a cross between farmer-improved Tuxpeño 1 (as the female line from Wentan village) and Jiahe white (as the male line from Zicheng village) in 2002. Both parents were selected by farmers, who have been involved in the whole improvement process. Xin Mo 1 is very drought resistant and yields, on average, 15% more than local varieties. It has not been registered. In 2003, it failed the registration trials in six provinces, which are required by the government before it can be formally recognized.
- **Zhong Mo 1**, an OPV, was derived from crosses of Xin Mo 1, Suwan 1 and Amarinto 966 (as the male line) in 2004. This variety was developed because PPB farmers wanted to improve Xin Mo 1, which is white, by creating a yellow variety, which would have a higher commercial value. This is the first cross between one parental line and an exotic line. Farmers have been involved from the F1 stage. It has not been registered.
- **Zhong Mo 2** was derived from a cross between Xin Mo 1 and Amarinto 9 in 2006. The objective was to produce a yellow variety and improve taste. Farmers and breeders from the GMRI worked closely together throughout the whole process. It has not been registered.
- **Guinuo 2006** is a hybrid waxy variety, also called Guangxi Wax 2006. It was produced by GMRI breeders using one line from a PPB project (in Duan county) in 2001. Since 2002, it has been tested, adapted and used for seed production in the PPB villages. Farmers have been involved in testing and adaptation since the F3 stage. It has not been registered.
- **Guizusong 2006**, an OPV, has strong lodging resistance and gives higher yields than local varieties. Farmers and breeders from the GMRI worked closely together throughout the whole process. It has not been registered.

The four OPVs have been tested and are cultivated in local communities without official release; the hybrid variety was registered under the name of a PPB breeder from the GMRI in 2004. Farmers have made various contributions to the development of these PPB varieties. They participated and contributed more in the case of the four OPVs than the hybrid variety, in terms of providing genetic materials, decision making, knowledge, labor and land. How this affects benefit sharing is discussed in a later section.

Seed production

Of the four OPV varieties listed above, Guinuo 2006 is the favorite among farmers and local communities, not only because of its exceptional taste, but also because of its market potential. In 2006, the PPB villages in Mashan and Long'An counties started local production of Guinuo 2006 seed. At first, the main difficulty for farmers was their lack of knowledge of hybrid seed production, but with help and technical support from GMRI breeders, they learned the basic skills and knowledge within two years. To manage the process better, farmers have set up a seed production group, who now produce seeds each season for their own use and to sell to neighboring villages.

To share the benefits of PPB products, we encouraged farmers and GMRI breeders to establish some agreements concerning the exchange of breeding material and seed production methods to enhance their collaborative relationship further. This sort of collaboration is still very new and requires time and effort by all parties to entrench the practice. It represents novel policymaking in practice and is being followed with interest by both the Ministry of Agriculture and the Ministry of Environmental Protection.

National policy and legal context regarding plant genetic resources, traditional knowledge and ABS issues

At the international level, China ratified the CBD in 1992 and, since then, has made good progress in conservation of plant genetic resources (PGR) and traditional knowledge and in ABS legislation. Pushed by economic interests and trade pressures, China joined UPOV in 1999 (the government signed the UPO 1978 version) and became a WTO member country in 2001. Under these competing international regimes, the domestic legislative framework has to balance interests of diverse sectors and groups to support sustainable development and social equity. In a country that is as large and diverse as China, this is not an easy task.

At the national level, there are no specific laws addressing ABS and traditional knowledge, but there are some related regulations and policies. Since 2006, the PPB team has been involved in a systematic policy review, analysis and discussions with policymakers and drafters. In parallel, the team opened up discussions with breeders and farmers from PPB trial villages and began experiments around protection of PGR and traditional knowledge and around ABS mechanisms for the PPB process and products. This method of trial and learning has gradually generated a certain common understanding in terms of policy influence in support of local level practices (Vernooy *et al.* 2007).

General review

There are two competing sides in legislation on PGR and traditional knowledge: ABS supporters and supporters of intellectual property rights (IPR). ABS supporters prefer to protect farmers' and communities' collective rights to PGR and traditional knowledge through such mechanisms as prior informed consent and mutually agreed terms. Meanwhile, IPR supporters work for an industry-oriented and individualized regime. The latter approach has been dominant so far, resulting in the threat that all PGR and traditional knowledge in China will be patented, and control over access, use and benefit sharing will be individualized. Between the two camps are those who prefer a more pragmatic solution: working for farmers' collective rights and benefits through what is called "soft" IPR, e.g. collective trademarks, copyright and geographic indications.

The following is a list of all key legislation and regulations related to PGR, ABS and traditional knowledge.

- *Regulation on Plant New Variety Protection*, established 1 October 1997 by the Ministry of Agriculture

  This regulation provides exclusive plant breeders' rights but leaves room for exemptions for research-oriented breeding and farmers' privileges for storing and using farm-saved seeds and propagating materials. Parts of the regulation are compatible with UPOV-91, such as the requirement on NDUS (a variety must be new, distinct, uniform and stable).

- *Seed Law*, enacted 1 December 2000 by National People's Congress of China

  This law protects breeders' benefits and opens up the domestic seed market to private entities. Under it, any company in compliance with the law can apply for a breeding license and a seed business license, and can conduct seed production and management within the permitted region.

  Article 2 mentions "conservation and use of genetic resources."

  Article 8 states that no institution or individual may possess or destroy genetic resources. The collection or farming of natural genetic resources is not allowed. Exemptions for scientific research or other special needs require approval from national or provincial authorities responsible for agriculture and forestry.

  Article 10 stipulates that the state has sovereignty over PGR; therefore, institutions and individuals cannot provide genetic resources to foreign institutions or individuals without approval from national authorities responsible for agriculture and forestry.

- *Rules of Seed Production and Operation License Systems*, established 26 February 2001 by the Ministry of Agriculture

  Institutions or individuals who conduct multiplication of main crop seeds need a seed-production license. Entities applying for a seed-production license should be well qualified. Institutions or individuals applying for a license for main crops must submit 5 million CNY (about US$760,500) as registration capital.

  Article 15 stipulates that registration capital of applicants who combine selection with production and operation must be more than 30 million CNY.

- *Rules for Multiplication of Main Food Crops*, established 26 February 2001 by the Ministry of Agriculture

  Entities applying for a seed-production license should be engaged in selection or artificial improvement; their products should be distinct from existing varieties, have relatively stable genetic characteristics, have uniform morphological and biological characteristics, and be appropriately named.

- *Rules for Management of Genetic Resources of Agricultural Crops*, established 8 July 2003 by the Ministry of Agriculture

  These rules are intended to strengthen protection, exchange and utilization of crop genetic resources. They contain a standardized working mechanism for identification, registration and conservation of agricultural genetic resources and regulate the process of importing genetic resources.

- *Outline of National Biological Species Resources Conservation and Utilization Plan*, established October 2007 by the State Council

  Priority action 6: Establish a legal system for the acquisition and benefit sharing of biological genetic resources and relevant traditional knowledge.

  Priority project 1: Define traditional knowledge, prepare a list of important traditional knowledge and establish a system to protect biological resources and traditional knowledge.

  Priority project 2: Establish a system requiring disclosure of the origin of genetic resources in patent applications.

  Priority project 3: Establish organizations for processing biological genetic resources and associated traditional knowledge, and a mechanism for information exchange.

  Priority project 4: Establish a database for the protection of biological genetic resources and traditional knowledge, and formulate a list of genetic resources and relevant traditional knowledge that should be protected.

- *Regulation Concerning Approval of Import and Export of Livestock Genetic Resources and International Collaborative Research Using Livestock Genetic Resources*, established in 2008 by the State Council

  Article 7: Any institution or organization that intends to export livestock genetic resources included on the protection list shall submit an application together with the following documents: (a) contracts or grant agreements governing such export and (b) documents that contain detailed benefit-sharing arrangements provided by the importer.

- *Outline of the National Intellectual Property Strategy*, established 5 June 2008 by the State Council

  Improve the protection, exploration and utilization of genetic resources, preventing their loss and abuse. Coordinate protection, exploration and utilization of genetic resources, and establish a reasonable mechanism of access to genetic resources and benefit sharing. Guarantee the rights of awareness and consent to the providers of genetic resources.

- *Science and Technology Progress Law* (2008)

  This law fosters the commercialization of public institutes/breeders by recognizing the intellectual property rights of public-sector agencies.

- *National Bio-diversity Protection Strategy and Action Plan*, established in 2009 by the State Council

  Strengthen studies of the genetic resources administration system at the national level, establish relevant policies that conform with national conditions. Besides the requirement of origin disclosure in patent applications, there is also a requirement for prior informed consent, so that the benefit can be shared, guarantee benefit to the resources providers and protect the interests of the vulnerable groups such as local communities and minorities. At the same time, it is recommended that: special laws be established for biological resources; a system of certification of the origin of biological resources be introduced; enforcement and custom inspections be strengthened; and illegal activities, such as smuggling of species, biological plagiarism and private export be curtailed.

The plan encourages investigation, registration and documentation of traditional knowledge and associated biological resources, including the establishment of *sui generis* regulations to protect traditional knowledge.

* *The Revised Patent Law*, enacted 1 October 2009 by the National People's Congress of China

Article 5 stipulates that inventions or innovations using genetic resources in a way violating laws or regulations related to access to and use of genetic resources shall not be granted patents.

Article 26 requires patent applicants to disclose the origin of genetic resources used in inventions or innovations.

Patent law in China can only be used for protecting innovations or inventions at microorganism and gene levels; not for protecting new plant and animal varieties. The latter can be protected by the *Regulation on Plant New Variety Protection*, enacted in 1997.

Constraints and opportunities arising from PPB and ABS policies and laws

In our review of legislation and policies concerning PGR, we identified some constraints to and opportunities for supporting and mainstreaming PPB and the conservation of agricultural biodiversity.

Constraints on PPB-related ABS

Regulation on Plant New Variety Protection (1997)

According to this law, plant breeders' rights are protected through an exclusive intellectual property regime, which conflicts with the collective holding of genetic resources. Farmers are not allowed to use germplasm freely from varieties protected under this law.

Seed Law (2000)

This law restricts farmer seed production from hybrids and allows for public breeders to commercialize their varieties (often produced by "borrowing" varieties from local farmers). This is in direct conflict with their public mandate to advance food security and rural development.

Science and Technology Progress Law (2008)

This has led to public goods (common pool resources) becoming privatized by the public sector.

Opportunities for PPB-related ABS

Rules for Management of Genetic Resources of Agricultural Crops (2003)

In these rules, the emphasis on identification, registration and conservation of genetic resources is closely linked with PPB actions.

Outline of National Biological Species Resources Conservation and Utilization Plan (2007)

The national database of biological resources and traditional knowledge can provide useful instruction for those wanting to register PGR. In regions rich in genetic resources and traditional knowledge, such as the southwest, there are opportunities to make use of this plan.

Outline of the National Intellectual Property Strategy (2008)

Some considerations related to ABS have been included in this strategy.

National Bio-diversity Protection Strategy and Action Plan (2009)

This plan emphasizes prior informed consent mechanisms and traditional knowledge protection, valuing farmers' rights to their genetic resources.

Revised Patent Law (2009)

In this law, the requirement to disclose the origin of genetic resources used in inventions or innovations has opened a window for change in terms of ABS.

Results

In general, the Regulation on Plant New Variety Protection, the Seed Law and the Science and Technology Progress Law have constrained PPB mainstreaming and related ABS implementation. They have caused conflicts in public institutes between their public service and market roles, and they are favoring privatization of public goods and government breeders' interests over farmers' interests. As a result, breeders have no incentive to carry out their public roles: serving farmers and managing natural resources sustainably. At the same time, the legislation and policy have harmed competiveness in the private sector.

The remaining laws and policies listed above offer some opportunities to support PPB and agricultural biodiversity conservation efforts at the local level.

The team has studied these opportunities in detail in light of our PPB practices, seeking to explore ABS mechanisms and continue policy discussions in collaboration with field and policy partners. The following section presents an analysis of these discussions and results.

Finding mechanisms for fair ABS

Along with the PPB activities carried out in the field, we addressed issues related to ABS in terms of both action research and development processes and the resulting products, the foremost of which are the newly developed maize varieties. (PPB is now also being applied to other crops in the southwest, including rice.) The team identified six stages in the PPB selection and breeding process that serve as entry points for ABS discussions and experimentation: germplasm registration, germplasm collection and conservation, germplasm use in hybrid breeding, germplasm use for OPV improvement, variety release and seed multiplication.

In the following summary, we focus on four questions at each of the six stages. What are the current barriers? Is there an alternative or innovative way to overcome these barriers? What is the best way to test the alternative or innovation? What are the challenges to moving forward?

Germplasm registration

Current barriers

Although formal public registration systems for germplasm exist at both provincial and national levels, current requirements focus on genetic and geographic information, omitting socioeconomic and cultural information about farmers and local communities where the material originated. In the formal system, the custodians of genetic resources are ignored or viewed as unimportant.

Alternative

The registration system should be amended to place more emphasis on farmers' and communities' rights and recognize their crucial roles in maintaining agricultural biodiversity in the field through both individual and collective efforts.

Action

The team has supported a germplasm registration system at the level of the three southwestern provinces. For the first time, farmers' information has been documented by breeding institutes during their germplasm collection missions to local communities and the role of farmers has been recognized. So far, 56 landraces from Guangxi, 46 from Guizhou and 71 from Yunnan have been systematically registered and analyzed in the lab, and more landraces have been documented during collection missions.

Challenges

Such a registration/documentation process is labor intensive and time consuming and requires a cross-disciplinary approach, i.e. combining natural science and social sciences during investigation.

Germplasm collection and conservation

Current barriers

Farmers continue to conserve and manage landraces in their local seed systems, which are under increasing pressure from market forces (e.g. both government agencies and private-sector businesses are staging campaigns to sell hybrid seeds). There are no incentives for farmers to maintain the local seed system. Breeders can obtain germplasm from farmers' fields for free (from interviews with farmers we know that breeders usually do not explain why they are collecting, let alone offer compensation for the material they collect). In practice, farmers have no control over their genetic resources and because their conservation efforts are not recognized, their awareness of the importance of conserving genetic diversity remains relatively poor.

Alternative

One mechanism for addressing this issue is to reward farmers for their conservation efforts when genetic resources are collected from them. However, it is not clear who should be compensated, given that genetic resources in the field are the "product" of many generations of (often collective) care and maintenance. Thus, there is need to tailor arrangements to the particular situation. For rare germplasm cultivated by only a few households, for example, compensation can be made directly to those households. For locally common germplasm cultivated by many households, compensation could be made to a farmers' group or a whole village to support public expenditures or to be used as a biodiversity fund, which could be accessed by local households in rotation, for example. The rationale for a biodiversity fund is that farmers who cultivate landraces from a given list (the varieties on the list should be as diverse as possible) could borrow from the fund. At the end of the season, those farmers would return the money together with seeds of adopted varieties, and the seeds could be provided to other households.

Action

The team has initiated a contract between breeders and farmers to regulate a fair and transparent germplasm access process. As mentioned above, the contract should reflect the distribution of the germplasm in the local community and result from communication and negotiations between breeders and communities.

In one of the PPB villages, a farmers' group has already used its community fund (which was set up with foreign-donor-supported project funds received by the CCAP) to provide monetary incentives for a few households to continue cultivating landraces that are disappearing. In the case of PPB, breeders make combinations with germplasm that has originated from farmers' fields with a claim on the source (as documented during the registration stage). Such recognition can be seen as a non-monetary benefit for farmers, and it will help clarify farmers' contributions in further breeding (see boxes, Examples 1 and 2).

Challenges

Because of the public value attached to genetic resources, incentives should be provided by the government and the public sector. This is a way to protect farmers' rights and enhance local genetic diversity. However, farmers' rights are rarely recognized by formal breeding institutes and regulatory systems. The team continues to discuss this issue with policymakers and argue for government support to set up biodiversity funds.

Germplasm use in hybrid breeding

Current barriers

Without transparency during registration and collection, germplasm use by outsiders seems to be a "black box." No one seems to be aware of the origin of the genetic resources, how they are maintained, what they are being used for and by whom, and how all of this matters to society at large. As a result, it is unclear how the benefits arising from the products of breeding processes should be shared. Breeders always play a dominant role in ABS negotiations, as plant breeders' rights are protected by the current regulatory framework, which does not recognize the role of farmers. In cases where breeders use local varieties for their crosses, they usually refuse to recognize farmers as their partner breeders.

Alternative

Under existing legislation, farmers' rights could be protected through prior agreement and formal contracts describing the benefits to be shared. The contributions of farmers could be determined in various ways. From our experience, such contracting processes could best be facilitated by a neutral third party. In addition to providing germplasm, farmers' efforts in the PPB process should also be reflected in the contract.

Action

After several rounds of discussions, the PPB process in Guangxi has become more transparent with clear indications of which of farmers' landraces are being used and what their contributions are to the crop improvement process. The formal institutes and farmers involved have decided to use formal contracts—the first ever signed ABS agreements for PPB in China. On 21 June 2010, two contracts

Figure 9.2 ABS agreement on conservation and improvement of maize and rice
        landraces in Guangxi

Source: Photo: Huang Jianqin

were signed by two breeding institutes, one policy institute (Example 1) and ten
villages in Nanning (Example 2). An English translation of the two agreements
appears below (see also Figure 9.2).

**Example 1. ABS agreement on conservation and improvement of maize
and rice landraces in Guangxi**

With the rapid loss of biodiversity, agricultural genetic resources have been
challenged. The mountainous area of southwest China is one of the biodiversity
hotspots in the world. Farmers living here are relatively far from the global market
and industrialized farming systems. To maintain genetic resources and improve
the livelihood of smallholder farmers in the southwest, the Participatory Action
Research (PAR) Program of the CCAP under the Chinese Academy of Science
(CAS) initiates this agreement for supporting farmers' in-situ landrace conservation
and improvement.

   The following items have been agreed to:

1. Each household is free to participate.
2. The program emphasizes landrace conservation and improvement with no
    exclusion of hybrid selection and adoption.
3. The program will provide technology and information support for participating
    farmers in collaboration with Guangxi Maize Research Institute and Guangxi
    Rice Research Institute.

4. To reduce the possible varietal and technical risks, the program will compensate according to the average yield of the local popular variety in the same season/year, at a similar location.
5. The program will provide incentives for pioneer practitioners who agree to evaluate the process and outputs of landrace conservation and improvement.
6. The costs of the program will be borne by the PAR Program, to set a Development Fund for interested communities. Compensation will jointly come from the Development Fund and the PAR Program; for communities without a Development Fund, compensation will be covered by the program.
7. Research institutes ought to subsidize farmers when collecting their landraces from project villages and provide source of collection when applying for national and provincial registration.
8. The particular institution to be set up for benefit sharing within the community is beyond the scope of this agreement, but the program will provide suggestions about developing concrete activities and setting up a benefit-sharing system in the community.
9. This agreement will be renewed every three years; it will enter into force after all parties have signed.

Example 2. ABS agreement on landrace use for hybrid maize and rice breeding in Guangxi

Facing the narrowed genetic base in hybrid breeding, recently, breeding institutes have put more emphasis on utilizing landraces and local varieties. Plant genetic resources distributed in southwestern China have provided the necessary guarantees for broadening the genetic base. The participatory maize breeding program in Guangxi has supported the process of utilizing landraces in hybrid breeding since 2000, through collaboration with public breeding institutes. For a long time, the contribution of farmers' landraces has been neglected technically and institutionally. It is recognized that, on the one hand, the process of hybrid breeding contains uncertainties in the process of germplasm selection, combination and recombination and, therefore, it is difficult to measure the specific contribution of landraces to the breeding process. It is also recognized, on the other hand, that there is lack of awareness and institutional willingness to share benefits with farmers after a new variety has been released. The Participatory Action Research (PAR) Program has facilitated public breeding institutes (the Guangxi Maize Research Institute (GMRI) and Guangxi Rice Research Institute (GRRI)) and project villages to reach an agreement on benefit sharing.

The following items have been agreed to:

1. As public breeding institutes, the GMRI and GRRI recognize the contribution of landraces collected from project villages in their hybrid breeding.
2. If farmer-selected/improved maize landraces are directly adopted as the parent line of hybrids, the farmers will get no more than 25% of the royalty.
3. If farmer-selected/improved rice landraces are directly adopted as the parent line of hybrids, the farmers will get no more than 25% of the royalty; if those

landraces can be directly approved as conventional varieties, farmers can get 100% of the royalty.

4. If released new varieties from public breeding institutes have used landraces directly or indirectly as the material for breeding, the villages and farmers from which these landraces were obtained have the right to a share of monetary benefits.

5. In the process of participatory plant breeding (PPB), farmer breeders can apply for plant breeders' rights jointly with formal breeders from public institutes. However, it is recognized that, before commercialization, no monetary benefit is usually generated.

6. Besides monetary benefit sharing, technical knowledge and experience exchange and sharing take place during the research process (at the project level). The program will strengthen collaboration between institutes and villages.

7. The particular institution to be set up for benefit sharing within the community is beyond the scope of this agreement, but the program will provide suggestions about developing concrete activities and setting up a benefit-sharing system in the community.

8. For PPB varieties, benefit sharing among the participants and naming will follow another agreement.

9. This agreement will be renewed every three years; it will enter into force after all parties have signed.

Contractors

Party A (institute representatives)
Three representatives from the Center for Chinese Agricultural Policy, Chinese Academy of Science, GMRI and GRRI
Party B (farmer representatives)
Ten farmer representatives from ten PPB trial villages in eight counties of Guangxi

Date and place agreed and signed

June 21, 2010
Nanning, Guangxi Province

Challenges

Designing ABS mechanisms that recognize farmers' rights fairly and effectively is not simply a matter of a moral obligation. These mechanisms must also be practical and be tested and evaluated. However, tracing and monitoring mechanisms and institutional arrangements concerning the flow of germplasm through public and private organizations and communities are nonexistent or poorly functioning. Most attention has focused on cross-country tracing and monitoring; at the national level, there has been little awareness of the importance of redistributing benefits across sectors, especially to farmers and their communities.

Germplasm use for OPV improvement

Current barriers

With commercialization of hybrid seeds (especially maize), both public breeding institutes and private seed companies have become purely profit-driven. Attention to OPV conservation and improvement has waned or almost disappeared. Although the numerous landraces in southwest China have been conserved by local smallholder farmers over many years, the almost total absence of research about and services for their conservation and improvement has led to their degradation on a large scale. Public-sector agencies have not shown much interest in rectifying this situation.

Alternative

A public-sector research program on landrace conservation and improvement should be set up as part of the working agenda of breeding institutes. Such efforts of breeders should also be recognized and evaluated in the institute's annual performance review (Figure 9.3).

Another option is to set up a registration system for OPVs (including landraces, traditional varieties and farmer-improved OPVs), in parallel with the new varieties protected by law. Within this system, the diversity of PGRs can be captured and the contribution of breeders (both farmer breeders and formal-sector breeders) can be recognized.

Action

Given the fact that four of five PPB varieties developed by the team in Guangxi are OPVs, PPB is, by nature, a public-value-oriented effort. PPB breeders (both farmer breeders and public-sector breeders) should take responsibility for germplasm conservation (with special attention to OPVs) and utilization (OPVs and their combinations). This role has been taken on seriously by the GMRI and GRRI.

Challenges

Public PPB breeders can obtain recognition and incentives by being involved in the research process, but given that this involvement is project-based, it is hard for them to receive professional recognition in the formal promotion system which focuses on yield increase through hybrid breeding. As a result, public breeders' work in and contribution to farmer-preferred PPB varieties does not count as work that contributes to their career (and opportunities for promotion) or the institute's performance as recognized by the government. A structural change at the institutional level is urgently required.

Figure 9.3 A PPB variety in the field (above) and harvested (below)
Source: Photos: Ronnie Vernooy

Variety release

Current barriers

PPB varieties include OPVs and hybrids. OPVs, which have lower market potential, are usually cultivated in local communities with no official release. In contrast, driven by the commercial value of hybrids, breeders are always seeking release of these varieties under their own names. This reflects the situation surrounding the five new PPB varieties developed by the team. It seems farmers have no opportunity to become formally recognized as breeders under current institutional arrangements.

Alternative

Given that three types of plant breeders are recognized in national plant variety protection law, i.e. duty breeder, collaborative breeder and individual breeder, PPB farmers can be identified as collaborative breeders together with breeders from formal institutes. According to the law, they could set up a form of benefit sharing through a formal contract.

Action

The hybrid PPB variety, Guinuo 2006, has been recognized as a PPB product, with acknowledgment of and respect for farmers' contribution of germplasm and breeding efforts. Community-based production of Guinuo 2006 seed has been supported by breeders since 2006, which is a form of benefit sharing. The recently designed PPB experiment in hybrid breeding has taken farmer breeders' rights into consideration from the beginning, and it will be followed with a contract (see item 3 in Example 2, above).

Challenges

Compared with breeders, farmers are usually vulnerable and marginalized. If farmers are to be recognized as collaborative breeders, our experience suggests that a facilitating agency act as a neutral third party; in our case, the CCAP's participatory action research group could be effective. Building trust is a main task for such a third party. Our ten years of commitment has been instrumental in this.

Seed multiplication

Current barriers

For OPVs, farmers can use their own saved seeds for subsequent seasons. Hybrids, however, are protected and registered under a breeder's name; thus, the community needs permission to acquire parent material for seed production. PPB hybrids fall into this category. Under the law, decisions on what seed to produce and how much are also under the control of breeders.

Alternative

If farmers were recognized as collaborative breeders, they would be able to participate in decisions about the transfer of intellectual property. A community-based seed production system would enhance farmers' seed systems. Such a system should be supported by public-sector agencies, in terms of technical support and services such as credit and information.

Action

PPB communities have been experimenting with community-based seed production since 2006 (Figure 9.4). To date, this experience indicates that farmers and local communities benefit from this kind of local economic enterprise, which is an important way to share benefits.

Challenges

It is difficult to scale up small community-based seed production systems. Farmers worry about the weather, proper techniques for seed treatment, seed storage and marketing (e.g. what is an appropriate price?). Farmers must rely on formal breeders to provide them with parent lines every year. Without legal recognition of farmers' rights, public breeders are reluctant to transfer or share their rights to large-scale seed production.

Additional policy experimentation and integration of results into policies and laws

In 2010, the team held a number of meetings and forums at the national level to discuss these issues with policy researchers and policy drafters from the Ministry of Agriculture and the Ministry of Environmental Protection. These encounters led to some concrete and fruitful suggestions that the research partners conduct additional or continued experiments and that the promising results begin to be integrated into policies and regulations addressing the six identified stages. These suggestions are presented below, followed by some concluding remarks concerning our ABS efforts.

Germplasm registration

The current national registration system should place more emphasis on farmer and community rights. The two-level germplasm registration method used by the team in southwest China, i.e. community registration and formal provincial

Figure 9.4 Seeds produced by a women's group on field (above) and harvested (below)
Source: Photos by Huang Kaijian (top) and Ronnie Vernooy (bottom)

registration, could be used as a model for reforming the national system. Some social, cultural and local indicators could be integrated into the existing formal registration system. In addition, each community should have its own registration system for local genetic resources, products and related traditional knowledge, beginning with maize and rice in the southwest as a pilot project. This work, which would benefit from technical support from the team, should use a systematic, cross-disciplinary approach, combining natural sciences and social sciences, integrating farmers' rights and the state's need for a sustainable seed system.

Germplasm collection and conservation

Given the nature of the land-use system (land is state owned), genetic resources are in the public domain. There is no compensation for farmers' efforts to conserve and manage genetic resources vital to the country's food security and other societal needs. Because of the value of genetic resources, incentives should be provided by the government and the public sector to farmers and their communities to carry on in-situ conservation and management. This would protect farmers' rights and enhance local genetic diversity and, just as important, enhance public awareness of the need to protect biodiversity and farmers' rights. "Conservation villages" could be selected, with a focus on different crops, plants and livestock, where dynamic systems of crop conservation and improvement are set up through the joint efforts of scientists and local farmers. Incentives should be given to these villages to encourage collective decision-making and collective action for management and conservation.

Farmers and their communities should also be compensated (in cash) when genetic resources are collected from them—by individuals, public-sector agencies or private business. The funds should be available to local communities for their own use relating to ABS issues. The agreement in Example 1, above, could be used as an initial model for further adaptation.

Germplasm use in hybrid breeding

The current registration and legislation systems do not adequately address farmers' rights and fair ABS. We suggest that the registration system be reformed first. Prior informed consent should be required, with an eye on ex ante agreement on potential benefit-sharing principles and practices between farmers' communities and breeding institutions. The contribution of farmers should be fully recognized—their contributions of germplasm, their traditional knowledge and their efforts in PPB should be acknowledged in the contract. Farmers' contributions could vary according to each situation. A general agreement (see Examples 1 and 2, above) could serve as a basis for elaboration of a more specific contract or subcontract, for a specific variety, for example. Such contracting processes should be facilitated, monitored and evaluated by a neutral third party, NGO or research institution and subsidized and supported by the government.

Germplasm use for OPV improvement

Although currently in China there is no legal recognition of collective breeding rights, there is the option of recognizing cooperative breeders. This would greatly benefit PPB farmers and allow them to participate formally in decisions concerning the transfer of intellectual property rights and the sharing of royalties and benefits. As there are three types of plant breeder rights in national plant protection law, i.e. duty breeder, collaborative breeder and individual breeder, PPB farmers could be identified as collaborative breeders together with breeders from formal institutes. According to the law, they can set up a benefit-sharing agreement by contract.

The formal seed system's recognition of farmers' roles and contributions is crucial to protecting farmers' rights and supporting their roles in biodiversity conservation. Given the value of landraces and OPVs, research in this area should be a priority on the working agendas of public-sector breeding institutes. Such efforts of breeders should also be recognized and valued in the annual performance evaluation of the institute and the personal promotion and reward system.

Variety release

In principle, both new hybrid varieties and improved OPVs are formally released at the regional and national levels. However, with the current institutional environment driven by market incentives, most released varieties are now hybrids because of their market value. OPVs are usually improved by farmers and cultivated in local communities with no official release. China could set up a new registration system for OPVs (including landraces, traditional varieties and farmer-improved OPVs), in parallel with the new varieties protected by current law. This would provide more incentives for both farmers and formal breeders working on OPVs. For PPB hybrids, farmers could apply together with formal breeders and, as co-breeders, receive a share of the benefits.

Seed multiplication

If the varietal release system recognized farmers as breeders or co-breeders of improved landraces and hybrids, fairer sharing would be possible. This is a strong incentive and invaluable support for community-based seed production of hybrids and high-quality landraces, which would enhance farmers' seed systems in terms of both biodiversity conservation and agricultural sustainability.

Conclusions

The field work in Guangxi and southwest China created a concrete entry point and platform for critical analysis of ABS issues. It also included experimenting with various ABS mechanisms with various stakeholders, with input from scientific disciplines and expertise from international, national, regional and local levels.

In general, we believe there is insufficient recognition of farmers' contributions to the maintenance and improvement of genetic resources and of their rights in terms of research, policy and law. The mainstreaming of PPB to link formal and informal systems and to ensure fair ABS could be the way forward. OPV improvement has been practiced by Chinese farmers since domestication of their crops and provides a good basis for synergies. PPB, and now so-called evolutionary breeding approaches (see Chapter 6), are needed to re-recognize and re-enhance farmers' roles, contributions and rights. Mainstreaming PPB in the current formal system would provide recognition and incentives for both public breeders and farmers for conservation, improvement, utilization and sharing of genetic resources in a fair and sustainable way. Some suggestions for action are as follows.

- Mainstream the PPB approach and practices in the formal public seed system to make it more farmer-oriented and environmentally friendly.
- Integrate ABS principles and mechanisms into the existing registration and legislation systems to promote fairer and more balanced collaborative seed systems.
- Set up a new registration system for OPVs (including landraces, traditional varieties and farmer-improved OPVs), in parallel with new-variety protection. Within this system, the diversity of genetic resources can be captured and the contribution of breeders (including farmer breeders and institutional breeders) can be recognized.
- Apply geographic indication to certain crops as a way to classify the area of origin for certain local high-quality seeds and derived products, for example, the waxy maize produced in some parts of Guangxi. All farmers within the area could benefit from collective production, and certification could be assigned and managed by a local farmer organization or a producer group with support from its village administration.

References

Ashby, J.A. (2009) The impact of participatory plant breeding. In S. Ceccarelli, E.P. Guimaraes and E. Weltzien (eds.), *Plant breeding and farmer participation*. Food and Agriculture Organization, Rome, Italy. pp. 649–71.

CCAP (Center for Chinese Agricultural Policy) (2004) Rural livelihood security and policy change—enhancing community-based crop development, natural resource management and farmer empowerment in Guangxi, SW China (research report). CCAP, Beijing, China.

—— (2008) Fair access and benefit sharing of genetic resources and traditional knowledge for rural livelihood security: exploring appropriate policies and laws in rapidly changing China. CCAP, Chinese Academy of Science, Beijing, China.

—— (2009) Survey report on the distribution of maize landraces in Guangxi, Yunnan and Guizhou. CCAP, Beijing, China.

Huang, J. (2003) Food security in China re-considered. Center for Chinese Agricultural Policy, Beijing, China (in Chinese).

Juan Zhang (2010) Poverty reduction in China in the last two decades. *21st Century Economic Report* Oct. 27.

Song, Y. (1998) "New" seeds in "old" China: impact study of CIMMYT's collaborative programme on maize breeding in Southwest China. PhD thesis, Wageningen University, Wageningen, Netherlands.

—— (2003) Linking the formal and informal systems for crop development and biodiversity enhancement. In *Conservation and sustainable use of agricultural biodiversity: a sourcebook.* CIP-UPWARD, Los Baños, Philippines. pp. 376–81.

Song, Y. and Jiggins, J. (2003) Women and maize breeding: the development of new seed systems in a marginal area of southwest China. In P.L. Howard (ed.), *Women and plant–gender relations in biodiversity management and conservation.* Zed Books, London and New York. pp. 273–88.

Song, Y. and Vernooy, R. (eds.) (2010a) *Seeds and synergies: innovating rural development in China.* Practical Action Publishing, Burton on Dunsmore, UK, and International Development Research Centre, Ottawa, Canada.

—— (2010b) Seeds of empowerment: action research in the context of feminization of agriculture in Southwest China. *Gender, Technology and Development* 14(1): 25–44.

Song, Y. and Zhang, L. (2004) Gender assessment report: impacts of IFAD's commitment to women in China 1995–2003 and insights for gender mainstreaming. International Fund for Agricultural Development, Beijing, China.

Song, Y., Zhang, L. and Vernooy, R. (2006) Empowering women farmers and strengthening the local seed system: action research in Guangxi, China. In R. Vernooy (ed.), *Social and gender analysis in natural resource management: learning studies and lessons from Asia.* Sage, New Delhi, India, China Agricultural Press, Beijing, China and International Development Research Centre, Ottawa, Canada. pp. 129–54. Available at: www.idrc.ca/openebooks/218-X (accessed 2 April 2011).

Song, Y., Zhang, S., Huang, K., Qin, L., Pan, Q. and Vernooy, R. (2006) Participatory plant breeding in Guangxi, southwest China. In C. Almekinders and J. Hardon (eds.), *Bringing farmers back into breeding: experiences with participatory plant breeding and challenges for institutionalisation.* Agromisa Special 5. Agromisa, Wageningen, Netherlands. pp. 80–6.

Song, Y., Zhang, S., Huang, K., Qin, L., Li, J. and Vernooy, R. (2010) Seeds of inspiration: breathing new life into the formal agricultural research and development system. In Y. Song and R. Vernooy (eds.), *Seeds and synergies: innovating rural development in China.* Practical Action Publishing, Bourton on Dunsmore, UK, and International Development Research Centre, Ottawa, Canada. pp. 47–64.

State Council Office of Poverty Alleviation (2007) *Annual poverty report, 2006.* State Council, Beijing, China.

UNDP (United Nations Development Programme) (2003) Overall report on China's accession to WTO: challenges for women in the agricultural and industrial sector. UNDP, United Nations Development Fund for Women, All-China Women's Federation, National Development and Reform Commission and Center for Chinese Agricultural Policy, Beijing, China.

Vernooy, R. and Song, Y. (2010) Changing rural development in China. In Y. Song and R. Vernooy (eds.), *Seeds and synergies: innovating rural development in China.* Practical Action Publishing, Burton on Dunsmore, UK, and International Development Research Centre, Ottawa, Canada. pp. 113–22.

Vernooy, R., Song, Y. and Li, J. (2007) Local agricultural innovation in China: ensuring a fair share of rights and benefits for farming communities. *Asia Tech Pacific Monitor* 24(2): 27–33.

Shihuang Zhang, Kaijian Hung and Lanqiu Qi (2010) PPT presentation at the National Maize Development Program meeting in September, Nanning, China. Center for Chinese Agricultural Policy, Beijing, China.

Zuo J. and Song Y. (2002) Women's experiences with "feminization of agriculture": insight from two village case studies in southwest China. Qinghua University, Beijing, China (in Chinese).

10 Cuba

The benefits of participation—
strengthening local seed systems

*Humberto Ríos Labrada and Ronnie Vernooy,
with Teresa D. Cruz Sardiñas*

New life for agriculture

Much like the economy at large, the Cuban agricultural sector is struggling to survive under difficult conditions. Farmers across the island, together with a number of young agricultural researchers, are realizing that necessity is the mother of invention. They are trying to breathe new life into the sector by revitalizing its heart: the seed systems that are the basis for production. In the process, new forms of participation and cooperation have emerged and, through these, new access and benefit-sharing (ABS) arrangements are evolving—not according to a predesigned plan, but as an expression of guiding principles that are informing the remaking of seed systems. These principles are based on a more flexible, open and dynamic view of how social change can be brought about.

The golden age of agricultural development in Cuba

During the heyday of the eastern European socialist countries, a centralized plant-breeding system was the model for Cuba's high-input agricultural crops, particularly the country's cash crops. Foreign varieties, hybrids, landraces and mutations were the principal sources of genetic material used for varietal development. At the end of the 10–12-year period typically required for varietal development for a specific crop, one or two varieties would be released for use in the entire country. Wide geographic adaptation was a characteristic encouraged by policymakers, and most governmental organizations involved in plant breeding (from research stations to university units) provided incentives to scientists to release varieties that could be used over a large area. In this sense, Cuba followed a modernization route similar to that of many other countries around the world.

In the 1980s, ambitious plant breeding programs were undertaken for sugar cane, roots and tubers, rice, tobacco, coffee, horticultural crops, pasture crops, grains, fibers and some fruit trees. These were carried out at 15 research institutes and their networks of experimental stations across the island.

Each new variety had to pass through a series of stages before it could be released. First, the research institutes submitted their results to the national scientific forum (Consejo Científico). This forum checked the scientific validity and, if it approved, sent its assessment on to an expert group of researchers, teachers and production unit directors. If the expert group approved the assessment, the results were then forwarded to the vice-minister of mixed crops, who would send them to the provincial delegations who would incorporate them into their production plans. In other words, producers, organized in cooperatives and state farms, were obliged to adopt them.

This procedure clearly represented a heavily top-down approach with no consultation with producers. Although during this period of socialist agricultural development, some researchers did visit farms, the research topics and problems they addressed had no input from farmers. The state controlled and managed the process and owned the results, including the actual varieties. Most benefits also ended up in the hands of the state. The salary of professionals involved in plant breeding was enough to sustain a Western lifestyle; they were well-respected in scientific and policy circles, and a key part of the modernization process steered by the state.

During these years, locally collected plant materials with useful features, such as disease resistance (crop diseases are a growing problem in the country), short growing cycles and good food qualities, were not prioritized by the formal breeding sector because of their relatively low yields under high-input conditions. However, they were maintained in one way or another by farmers across the country. Later, when high-input agriculture would dramatically collapse, the importance of these varieties would be recognized, and the country would owe gratitude to the farmers who painstakingly maintained them under difficult conditions. It was then that questions about who should have access to and benefit from these plant materials would arise.

A system in crisis

With the disintegration of the Union of Soviet Socialist Republics in 1989, the Cuban agricultural sector had to cope with a drastic reduction in imports and foreign trade support, which resulted in a gradual shift toward more self-sufficient and "rational" forms of production, i.e. more adapted to local conditions. Many remarkable technical and social transformations occurred in response to this new situation and challenge. Practical experiments—some small, some larger—all over the country emerged to give new life to the agricultural system and a way to survive under severe stress.

In the 1980s, 87% of Cuba's external trade had benefited from preferential price agreements, especially with countries of the former Soviet bloc. The country imported 95% of its fertilizer and herbicide and owned one tractor for every 125 ha of farm land. After the collapse of the socialist bloc, foreign purchase capacity was reduced from US$8,100 million in 1989 to US$1,700 million in 1993, greatly reducing the country's ability to buy agricultural inputs and keep industrial agriculture running. Many tractors came to a full stop. Many crops had to be grown with heavily reduced amounts of fertilizer and other inputs or with none at all.

To address the crisis and reduce the negative impact on the national economy, the Cuban government implemented changes in all sectors. During the early

1990s, a series of severe social and economic measures were taken to maintain the basic social guarantees of the government; at the same time, a more thorough reconstruction of the Cuban economy was initiated. Cuba, thus, undertook a dramatic change in its farming system, moving from being the largest agrochemical consumer in Latin America to a system using few external inputs—and in three years. This process also forced professional scientists to alter their living standard as they were now earning the equivalent of US$3 a month. The crisis led to considerable migration: people from the cities moved to the rural areas and a critical mass of people left for the United States, Europe, Latin America and elsewhere. Suddenly, Cubans, including many scientists, appeared all over the world.

During the same years, civil society underwent many changes. NGOs appeared and filled various gaps, and many small private businesses sprang up across the country (some of them disappeared again quickly, but others persevered). The rigid environment so common in previous decades changed drastically. Cubans began to realize that, although the revolution in 1959 had freed them from the United States, the many socioeconomic advantages they had acquired depended strongly on the socialist countries and were actually very fragile. The Cuban revolution achieved unexpectedly high levels of public health, food security and education, but it was economically vulnerable. New practices were urgently needed to keep the population alive.

Cuban plant breeders, most of them still clinging to a Green Revolution paradigm, were slow to adapt to the new situation: lack of fuel, unavailable spare parts and low-input agriculture. Because of the financial crisis, research institutions faced various constraints, such as lack of access to technological packages, the need to maintain important genetic resource collections, energy blackouts, inability to renew seed supplies and a decrease in the number of international programs that had supported Cuban research institutions in the 1990s. The national seed supply system urgently needed to expand and innovate, but lacked the financial resources to do so. In the 1990s, its seed production capacity for maize and beans had fallen by 50%.

Even though professional plant breeders faced a difficult economic situation and were offered few incentives, they continued to pursue top-down approaches and maintained a rigid reductionist perspective. They believed that the best solution for all these problems in agriculture and plant breeding was "simple" technology substitution. However, more fundamental changes were required.

With agriculture relying more and more on limited inputs, the production of seeds of the basic staples of the Cuban diet became a major issue in many parts of the country. Seeds from all over the country had provided a basis for plant breeders to select commercial genotypes during the industrial agriculture period. However, relatively little attention had been paid to the informal seed management systems. Researchers soon found out that, unfortunately, in these systems, much genetic variability had been eroded.

Taking advantage of the opportunity presented by the economic crisis, a small group of professionals set out to address this situation. In 1999, they designed a pilot project aimed at developing participatory seed production, improvement and distribution practices. This program, which has evolved over the years into a much larger, national level initiative, uses a variety of tools, including seed fairs and participatory variety selection, as strategies for seed diversification, yield improvement and the dynamic maintenance of genetic diversity in Cuba (Vernooy and Stanley 2003; Ríos Labrada 2006). Those involved in the process (farmers, plant breeders, technicians) developed new attitudes toward farmers' traditional knowledge and skills, their capacity for experimentation, their ABS principles and practices, and their participation and cooperation in research and development efforts, as well as in societal change more broadly.

Start of a new beginning

In 1999, 15 farmers and eight scientists (agronomists, biologists, biochemists and one sociologist) from the National Institute of Agriculture Sciences (INCA), the Agrarian University of Havana and the Sociological and Psychological Research Center in Havana pioneered "participatory seed diffusion" (PSD), the term the team used to summarize their innovative efforts. The initiative was supported by the International Development Research Centre and the Participatory Mesoamerican Program (funded by the Development Fund of Norway) and later by the Canadian International Development Agency, the Swiss Agency for Development and Cooperation, ACSUR-Las Segovias and the German organization Agroaction, allowing it to be scaled up to other regions and policy levels. The ministers for higher education and for science, technology and the environment have been supportive from the earliest stages. In 2010, after nine years of experimental implementation, more than 50,000 farmers were involved in PSD and more than 20 institutions were disseminating the approach all over the island. The number of farmers involved has continued to grow.

Paradigm shift toward participatory seed diffusion

Inspired by the practice of participatory plant breeding, which was gaining ground at the end of the 1990s in a number of countries around the world (see other chapters, and Vernooy and Stanley 2003), the initiative introduced the concept and practice of PSD as a way to integrate diversity seed fairs with farmer experimentation. Seed diversity fairs are events where plant breeders, farmers and extension agents have free access to diverse varieties of one or more crops. Varieties from formal and informal seed systems are sown under the usual cultural conditions in the target environment, then farmers are given free access to all the seeds and can choose the varieties they want in the fields. They take seeds from the selected varieties (or materials under development) back home for further experimentation.

Farmers have taken up this offer with great enthusiasm. Across the country, they have planted selected seeds and discussed the ensuing results with various stakeholders. Through this learning-by-doing process, farmers have adopted a sound research logic to obtain scientific results, and professional researchers have changed from technological decision-makers to facilitators of farmers' welfare—a major sea change.

In contrast to the former centralized model, PSD is based on individual farmers organized in agricultural production cooperatives or farmer experimenter groups, who test and then distribute varieties of high interest to the community. This is done on the basis of the traditional notion of reciprocity. Starting with the introduction of genetic diversity in a number of sites, over time seed diversity nodes are being developed and, through a "chain-reaction" process, diversity is increasing exponentially with increased farmer participation.

Once farmers have seen the favorable results of these experiments, they have tended to organize themselves into research groups. Each diversity node or nucleus promotes knowledge, social organization and entrepreneurial activities characterized by intense genetic flows, value-added efforts and continuing discussion around local innovation more broadly. Over time, as more and more farmers and others have joined the process and results have become evident, questions of recognition and ABS have become a core part of these discussions. In the following sections, which describe the evolution of the program, we focus on this issue.

Before doing so, we summarize the current legal framework surrounding access to and use of genetic resources. Although this framework has been in place for some time, it is general in character and, in practice, its significance has been very limited.

The legal framework for access to genetic resources in Cuba

Teresa D. Cruz Sardiñas

The first ABS regulations in Cuba were implemented in 1996, soon after the country ratified the CBD on 9 March 1994. ABS provisions regarding genetic resources are included in various legal instruments—laws, decree-laws and resolutions—creating a certain synergy between them; however, complementarity is missing in some cases. Responsibility for biological resources is shared by the Ministries of Science, Technology and Environment and Agriculture and Food Industry. They have jurisdiction over flora, fauna and marine resources, and deal most directly with ABS issues.

The Constitution

The Constitution of the Republic of Cuba declares the sovereignty of the state over living and nonrenewable natural resources, national waters, the seabed, and the subsoil of the maritime economic zone, as established by law and according to international practice (Article 11(c)). Article 15(a) establishes as socialist state property "natural resources both living and nonliving within the maritime economic zone of the Republic, forests, inland waters." Although there is no specific reference to genetic resources, sovereignty and state property principles are applicable, as, according to the CBD (2008), genetic resources are biological resources.

The Environment Law

Passed on 11 July 1997, Law 81, the Environment Law (Cuba 1997) is the framework within which Cuban environmental principles, objectives and management guidelines are established. With regard to ABS, Law 81 contains two relevant articles in Chapter 2, "Protection and use of the biological diversity," in the Sixth Title "Specific Areas of Environmental Protection." Article 85 establishes a special protection regime for genetic resources, which includes the establishment of strict regulatory, control, management and protection mechanisms to guarantee their conservation and appropriate use. Article 88, on the other hand, empowers the Ministry of Science, Technology and Environment to undertake special protection of genetic-resource-rich ecosystems and natural habitats. It calls for the development of measures to maintain species and their evolutionary processes in their natural environments. The ministry is also empowered to "establish or propose, as appropriate, the necessary strategies and regulations to guarantee a fair and equitable sharing of benefits arising from utilization of genetic resources." This is the first, most direct reference to genetic resources and the benefit-sharing concept, albeit in a very general context and subject to further regulation.

Forest Law

The Ministry of Agriculture, in coordination with the Ministry of Science, Technology and Environment, is responsible for enacting regulations regarding the reproduction, management, and conservation of forest species, including specimens and specific genes (Cuba 1998).

Biosafety Law

Decree-Law 190 (Cuba 1999a), on biosafety, establishes a general framework regarding the use, research, testing, production, import and export of biological agents and their products, organisms and fragments with genetic information. It obliges state entities, in particular those in charge of research facilities and release areas, to coordinate (through permits) with the Ministry of Science, Technology and Environment when undertaking activities related to the use of biological agents and their products.

Environmental Liabity Law

Decree-Law 200 (Cuba 1999b) establishes the legal framework for environmental liability and damage to the environment. This liability applies to both national and foreign natural persons and legal persons. Criminal and illegal behavior includes accessing biodiversity resources without appropriate authorization. The penalty is a fine of 250 pesos for natural persons and 5,000 pesos for legal persons, plus other measures that may include seizure of collected materials or the prohibition of certain activities.

Resolution 111/96

This resolution of the Ministry of Science, Technology and Environment (Cuba 1996) is the main legal instrument governing access to genetic resources and distribution of benefits. It refers mainly to the need for researchers (with commercial and non-commercial interests) to share the benefits derived from access to and use of Cuban genetic resources through prior informed consent and mutually agreed terms reached with the state authority. As a general rule, research in Cuba should be carried out by Cuban researchers. However, the resolution also extends to access to biological resources through a system of permits and benefit-sharing provisions granted and defined by the ministry as the competent authority. Although these rules are applicable to plant breeding and improvement of seeds (biological and genetic resources at the same time), implementation of the resolution has been weak.

The first seed diversity fair

In 1999, the first seed diversity fair was held at INCA to disseminate maize seeds adapted to low-input agriculture. At the fair, professional breeders provided invited farmers (men and women from the cooperative sector and others) with access to a wide range of varieties bred in the formal and informal seed systems. Eighty farmers from regions of high-input production, along with formal-sector maize breeders, social scientists from the National Agricultural Research System and representatives from the National Small-Farmer Association and the former Cuban Association of Organic Agriculture attended the fair.

Some months before the fair, two INCA breeders undertook maize seed collection missions to farming communities in the province of Pinar del Rio and Santa Catalina in Havana province. Selecting for hardiness under low-input conditions, they collected 66 landraces, including ten from communities in Havana province. In addition, four commercial varieties were contributed by other research institutes. All the seeds were provided freely, without any regard to recognition or ABS at this time.

Seeds were planted in December on an experimental plot at INCA. Each of the 70 lines was sown in three rows, and wide border strips were sown with a mixture of different lines. The experimental plot was irrigated only once and no fertilizer or pest control treatment was used.

At the fair, farmers were invited to inspect the maize experimental plot and to examine the cobs on all the maize lines. They were then each invited to select the five they preferred. Seeds from these lines would later be given to the farmers for experimentation, free of charge. Short questionnaires were used to gather information on the farmers' evaluations of the varieties they chose, and the results were discussed in plenary. The farmers also identified the main problems associated with seed management and use: poor seed quality, unavailability of seed and the incidence of pests and diseases. Availability of training and extension services, seed exchanges and access to inputs were considered less problematic.

Farmers showed an immediate preference for varieties in the mixed borders, as these showed a better response to low-input conditions than the plants in the mono-varietal rows. This led researchers to conclude that they would have to determine how to change their practice of maintaining varieties through strict isolation as advocated in the formal seed system. Farmers not only looked at yield, but also valued such factors as plant height, stalk size, number of cobs and number and position of leaves. This was another indication that the plant breeders should start thinking about alternative breeding strategies.

The response to this new participatory approach was positive, given that farmers were accustomed to a more top-down management procedure. Farmers had quickly and easily selected from the 70 lines on show, and a large range of new seed lines had been extended to them. The plant breeders involved in PSD saw the need to refocus seed management to improve yields and cob quality under low-input conditions. They believed that stimulating the flow of genetic resources could improve crop performance in a broad sense. Encouraged by the results of the first seed fair, they continued along this line and increased their efforts. They also became more conscious about the need to deal with recognition and ABS questions, although, at this stage, no action was taken.

Embracing diversity: farmers breeding maize at the local level

After the fair, the research team looked at farmers' maize populations and were surprised to discover an interesting genetic mosaic. The maize population of one Havana farmer who had participated in the seed fair was found to be composed of varieties with different origins: one commercial variety from the formal seed sector, five half-sibling varieties of a landrace from La Palma (a neighboring province) and four half-sibling varieties of a landrace from Catalina de Guines (a neighboring municipality in the same province). Later, the same farmer did a bulk planting of all his planting material then selected the best 1,500–2,000 plants according to cob size, plant cob height and husk covering, over three growing cycles. Later, at a seed fair organized by the farmer's cooperative, the combined population was sowed along with 38 landraces conserved by the Fundamental Research Institute (INIFAT) gene bank, 56 half-sibling varieties of landraces maintained by INCA, four commercial varieties and the male parent of a popular hybrid.

Subsequently, two mass selection cycles were carried out on the combined population. Gradually, this new seed pool, under farmer management, was used to increase maize production and for diffusion to cooperatives; as a result, the area intercropped with maize increased over the years. Maize rose from being one of the most neglected crops in the cooperative to the third most profitable crop. Currently, this population, called Felo variety, is in the seed multiplication and continued selection stage, having gained recognition of municipal stakeholders. Felo has also been registered as an official variety in Cuba.

This is a remarkable sequence of events, central in which is the recognition by professional plant breeders, other scientists, leaders and policymakers alike that farmers are capable of contributing to crop improvement as plant breeders in their own right and that they deserve to be known as such. No stakeholder has limited access to the new variety or tried to block the generation and sharing of benefits (the cooperative has become a seed producer). The process leading to Felo created new understanding, attitudes and behavior in terms of recognition and ABS, facilitated by farmers and breeders alike.

Felo and his cooperative were pleased with this success and empowered by the recognition they received as excellent plant breeders. However, the collaborative approach Felo and his colleagues used was heavily criticized by national and international conventional plant breeders. The main criticism centered on the fact that Felo allowed different maize varieties to be freely cross-pollinated in the field, contrary to the conventional model which entails recombination during the first stage of the breeding program and isolation of a population with the desired characteristics. This model is part of the conventional (Green Revolution) paradigm, in which homogenization, industrialization and decisions about what is "best" are made by those at the top. The Felo experience shows that alternatives, based on the potential of diversity and bottom-up decision-making, are feasible. Felo has increased yield, diversity and culinary quality in Cuba's maize gene pool by using his own selection and experimentation criteria. Now, hundreds of farmers are replicating that approach.

The recognition of Felo variety as a collaborative initiative between Felo's cooperative and scientists from the National Institute of Agricultural Science may be the first evidence in Cuba that improved varieties can be developed by farmers and researchers together. An increasing number of farmers now want to register varieties of maize and are demanding the establishment of a way for them to sell seeds.

Reframing a poor question: which bean variety is the best?

It is common to hear plant breeders say, "This is the best variety and that one is the worst." However, providing access to seed diversity for farmers goes beyond this conventional and restricted thinking. It acknowledges human diversity and tries to "transplant" this basic feature to the process of plant selection and improvement.

In the case of common bean, a self-pollinated crop, the PSD program has been working mainly with released varieties and landraces, using a nonsegregating population. Farmers have access to up to 124 varieties of beans from various sources grown under low-input conditions at INCA's experimental station. Each variety is sown in a small plot, and participants may select up to five varieties to take home and test on their own farms.

At a diversity fair for bean seeds, participating farmers came from different biophysical and socioeconomic contexts; both marginal and industrial farming systems were represented by 42 farmers. Also present were scientists from the National Agriculture Research System, NGO staff and functionaries and technicians from the Ministry of Agriculture.

During the fair, the program team asked men and women farmers to choose varieties separately, and administered a questionnaire to determine whether there were differences in selection criteria according to sex. Sixty bean varieties were cooked and participants were grouped in small teams of three men and three women to evaluate ten varieties each; a questionnaire on cooking qualities was also completed by participants with help from team members. The results were then analyzed by the team.

Male farmers voted for varieties with high yield and associated characteristics, such as number of pods per plant, pod size and disease resistance. Female participants voted for varieties with large pods and looked at grain size, shape and color, i.e. their criteria seemed to be more related to culinary properties. In the cooking test, men noted that more than 80% of the varieties tested were of good quality, whereas women were more rigorous. This is not surprising given that Cuban women do most of the cooking, but in conventional bean breeding, culinary aspects (and related features, such as cooking quality) are seldom considered.

Most farmer participants associated grain color with variety and were interested to see differences among beans of the same color; they especially commented on the degree of variability in disease resistance within the same color group.

In principle, the selection exercise was run on an individual basis; however, some farmers collectively decided to choose a wide range, as they wanted to test a range of varieties in their region. They were keen to organize a seed diversity fair exercise in their own communities. During the selection exercise in the field, the project team noted that none of the farmer participants had had access to such broad genetic diversity before.

After the bean seed fair, the program's mission was to compare and release varieties grown according to farmers' traditional farming systems. Workshops on experimental design were held at the community level and experimenter farmers' networks began to grow a large number of bean varieties (some farmers planted more than 100). Confronted with such bean diversity, scientists were overwhelmed; no one had expected genetic diversity to be of such importance to farmers. The question of which is the best bean was never heard.

The main interest of farmers was to be able to select among the wide range of varieties of maize and beans according to their own criteria and free from formal restrictions. Numerous varieties conserved in the gene bank (and tried out at the fairs) displayed good performance even though some had been removed from official varieties lists. The spirit of experimentation, the opportunity for more productive options and the importance of gender-based differences detected in the first participatory seed selection exercises in Cuba inspired farmers, scientists and other stakeholders to explore PSD further in Cuba. Consequently, the PSD team started to collect seeds from different sources and promote diversity seed fairs and farmers' experimentation in different regions, celebrating diversity and the freedom to use, assess and disseminate its richness. They were creating new expressions of ABS.

One of the main impacts was the increasing number of crops handled in PSD over time. The approach began with maize and beans, but six years later more than 20 crops were in the hands of the farmers' experimental network, and crop yields and the number of varieties managed by farmers increased significantly (Table 10.1).

Table 10.1 Crop yield and number of varieties in La Palma, Pinar de Rio province over four years of PSD

	Crop yield (t/ha)		No. of varieties	
	Before PSD	*After four years of PSD*	*Before PSD*	*After four years of PSD*
Tomato	8.0	12.0	3	42
Maize	1.4	2.4	4	52
Bean	0.4	1.4	5	200
Rice	1.9	3.8	6	45

Seed diversity fairs and farmers' experimentation gave farmers the right to choose varieties and make decisions about growing strategies on-farm—key elements in the participatory approach for many crops in Cuba. However, there are some differences between crop types in terms of the recognition of farmers as creators of new varieties. For example, maize is a cross-pollinated crop, and the many genetic combinations available at seed diversity fairs and through farmers' own experiments allowed farmers to construct their own gene pool and gain recognition as official variety creators. In contrast, beans are self-pollinated crops and PSD has focused on dissemination of diverse improved varieties or landraces historically managed by communities; here, the role of farmers as new actors in plant breeding is not so evident.

Although hundreds of varieties have been produced by farmers and freely disseminated by them, it has not yet been feasible, given policy and legal conditions, to give farmers the chance to produce official varieties of seed, beans in particular. Under current agricultural policy, this role remains with conventional research institutions.

Decentralized seed production

Over the years, the research team has noted some differences between PSD and conventional plant breeding in terms of seed production concepts. A defining characteristic of PSD is the integration within the household or community of genetic resource conservation, plant breeding, seed production, crop production and food consumption. In contrast, in convenional plant breeding, these functions are institutionalized, specialized and separate. For PSD in both marginal and industrial environments, the tendency has been to maintain diversity as much as possible. The rationale is, "We need to keep various options because who knows how hard the next season could be" (from an exchange between Cuban and Syrian farmers recorded in the field by H. Rios, 2005). Through PSD, farmers reinforce local seed production and exchange for further experimentation in the next cropping season or simply for culinary testing. They also use seeds for promotion or trade them in exchange for other products. Curiously, some farmers who never before produced seeds are now selling seeds to other farmers or to the state seed company.

The formal scheme of releasing certified seeds (for adoption by farmers) has broken down. With PSD, as with other PPB methods, farmers adopt varieties through experimentation and immediately release their best options. Thus, farmers are participating in the whole process of seed selection. Seed production is now an integrated process in the hands of farmers who decide the varieties or species that they want to multiply and disseminate.

Scaling up, changing mindsets and opening doors

The promising results achieved in the first years prompted the team and partners to amplify the pilot experience. They were eager to find out how PSD could be adapted to other parts of the country with different biophysical and socioeconomic features, but were worried about the risk of adopting a top-down approach in disseminating their work (and thus regressing to conventional practice). How could they upscale PSD while maintaining a dynamic, open process, based on the collective efforts of various local stakeholders?

In terms of what to upscale, the team decided on the following:

- The capacity to examine genetic diversity, with identification by local stakeholders of their own intervention entry points and facilitation of an enabling institutional environment
- The idea of seed diversity fairs and farmers' experimentation with various crops to develop varietal demand, linked to enhancement of farmers' participation in generating benefits
- The encouragement of conventional research institutions, particularly universities, to work directly with farmers in facilitating access to genetic diversity, farmers' experimentation and dissemination of seeds.

The team set out to discuss the idea of PSD with a wider range of stakeholders, and received constructive reactions from government, local universities, local research stations, civil society and farmers. At various sites across the country, individuals from various organizations joined forces to build teams, plan new activities and start work. Local organizations have been extremely cooperative in supporting the process, effectively moving the process forward and gradually becoming PSD leaders in their regions.

In the process, some "old" beliefs have been left behind. Although many participants had always regarded farmers as knowledgeable and having a right to contribute to policymaking, they also held the conviction that development could not take place without government subsidies. The PSD program experience demonstrates that farmers who are making decisions about the kind of agriculture they want (one form of innovation) can generate enough benefits to develop a region without help or with limited subsidies and other external resources.

In summary, many local stakeholders began to see farmers not as a burden, but as a solution to strengthen local innovation systems in agriculture.

The experience indicates that farmers who maintain more diversity and participate in dynamic seed exchanges gain considerable social recognition and also increase profits. The capacity of farmers to experiment seems to be an important element in a successful family business. Everywhere seed diversity fairs have been held, farmers have shown great interest in introducing more genetic diversity into their own farms. PSD is an attractive initiative not only for farmers but also for technicians, researchers, functionaries, politicians and policymakers, who are learning about the opportunities offered by genetic diversity for revitalizing cropping systems (and using fewer agrochemicals). They are also becoming aware of the importance of local knowledge for innovation.

The PSD team has come to understand that what has been scaled out is a pedagogical process more than a technological one. A wide range of stakeholders has learned how to produce and share benefits through meaningful and effective participation in Cuban society. In this process, many obstacles need to be overcome to let people develop their own initiatives. Although the Cuban government has built the capacity to innovate, the results have often not been satisfactory. In practice, PSD is an alternative that supports real stakeholder integration and more concrete forms of participation, and it is producing clear benefits.

Through its results in terms of yield, diversity and the enthusiam it has generated, PSD has been an interesting way to introduce the participation concept into the Cuban agricultural context. It has illustrated the potential of collaborative efforts of farmers and scientists in improving farming systems. Indeed, PSD reorients the plant breeding concept in terms of benefit sharing and generation of more development options for both farmers and scientists.

References

CBD (Convention on Biological Diversity) (2008) Report of the meeting on the group of legal and technical experts on concepts, terms, working definitions and sectoral approaches. Secretariat of the CBD, Montréal, Canada. UNEP/CBD/WG-ABS/7/2. Available at: www.cbd.int/doc/meetings/abs/abswg-07/official/abswg-07-02-en.pdf (accessed 3 March 2011).

Cuba, Republic of (1996) Resolution 111/96 of the Ministry of Science, Technology and Environment. *Gaceta oficial de la República de Cuba, edición ordinaria* XCIV(40): 631. Republic of Cuba, Havana, Cuba.

—— (1997) Law 81. *Gaceta oficial de la República de Cuba, edición extraordinaria* XCV(7): 47. Republic of Cuba, Havana, Cuba.

—— (1998) Law 85. *Gaceta oficial de la República De Cuba, edición ordinaria* XCVI(46): 773. Republic of Cuba, Havana, Cuba.

—— (1999a) Decree-Law 190. *Gaceta oficial de la República de Cuba, edición ordinaria* XCVII(7): 114. Republic of Cuba, Havana, Cuba.

—— (1999b) Decree-Law 200. *Gaceta oficial de la República de Cuba, edición ordinaria* XCVII(83): 1339. Republic of Cuba, Havana, Cuba.

Ríos Labrada, H. (ed.) (2006) *Fitomejoramiento participativo en Cuba: los agricultores mejoran cultivos.* Ediciones INCA, La Habana.

Vernooy, R. and Stanley, B. (2003) *Farmers and researchers reshape Cuba's agriculture* (case study). International Development Research Centre, Ottawa, Canada. Available at: http://publicwebsite.idrc.ca/EN/Documents/10558774410Seeds_3_Cuba.pdf (accessed 3 March 2011).

11 Nepal

Innovative mechanisms for putting farmers' rights into practice

Bikash Paudel, Kamalesh Adhikari,
Pitambar Shrestha and Bir Bahadur Tamang

Innovation at work

Until the beginning of the 1990s, ABS was an unknown concept in Nepal. There was no recognition that an ABS regime could form the basis for the protection of the rights of local, indigenous and farming communities over genetic resources and associated traditional knowledge. Notwithstanding a substantial international focus on these issues and their strong relevance for biodiversity-rich countries such as Nepal, Nepal's constitution of 1990 did not address this area. It was only after Nepal became a party to the CBD in February 1993 that the government and some NGOs began to discuss the importance of mainstreaming ABS issues in national policies. Similarly, following the country's engagement in the FAO Commission on Plant Genetic Resources, national-level discussions were held to undertake initiatives for the conservation, management and use of plant genetic resources for food and agriculture (PGRFA) and, in the process, seek options to address farmers' concerns with regard to PGRFA and associated traditional knowledge.

In this case study, we describe innovative research and development efforts to give concrete meaning to the concept of ABS in practice and to create an enabling policy and legal environment in favor of the diversity, both biological and sociocultural, on which Nepal depends. These efforts were led by two NGOs— Local Initiatives for Biodiversity, Research and Development (LI-BIRD) and South Asia Watch on Trade, Economics and Environment (SAWTEE)—who jointly executed a project named "Promoting innovative mechanisms for implementing farmers' rights through fair access to genetic resources and benefit sharing regime in Nepal" with financial support from the IDRC.

Diversity richness

Nepal is a mountainous country with an area of 147,181 km^2. It rises from 72 m above sea level on the northern rim of the Gangetic plain to some 90 peaks beyond the perpetual snow line over 7,000 m high. In addition to the continuum from tropical warmth to cold alpine areas comparable to polar regions, average annual precipitation varies from as little as 160 mm in the rain-shadow north of the Himalayas to as much as 5,500 mm on windward slopes.

Nepal is also diverse in social and economic characteristics, as the country is home to various races, tribes, languages, dialects, cultures and religions. There are about 59 indigenous ethnic groups who speak 22 languages and 96 dialects. Nepal's population is estimated at 26 million with an annual growth rate of 2.25% (CBS 2003, MOF 2008).

The extreme variations in altitude, topography, climatic conditions, sociocultural composition and farming practices have resulted in immense diversity in natural flora and fauna as well as cultivated crops. Comprising less than 0.1% of the earth's land mass, Nepal harbors 10% of all birds (862 species), 4% of mammals (181 species), 1.53% of reptiles (143 species of reptiles and amphibians), 6% of bryophytes (687 species of algae), 3% of pteridophytes (1,500 species of fungi and 465 species of lichens) and 2% of flowering plants (about 7,000 species) and, hence, ranks 31st in the world in terms of biodiversity. In addition, the country is home to about 200 species of commercially important medicinal and aromatic plants, 5,000 species of insects, 185 species of fish, 400 species of agrohorticultural crops, 60 species of wild edible fruits and 300 species of orchids (MoFSC 2002, Gautam 2008).

Paving the way for ABS

LI-BIRD and SAWTEE are two NGOs active in the field of biodiversity conservation and genetic resources policymaking. Currently, they are aiming to contribute to the institutionalization of community-level mechanisms for the effective implementation of an ABS regime suitable to the Nepalese context with particular interest in the sustainable use of genetic resources and associated traditional knowledge. The expectation is that these innovative mechanisms will effectively give meaning to the concept of farmers' rights. In terms of research objectives, the two NGOs are working together to:

• assess the appropriateness of policy and legal instruments to implement farmers' rights relevant to access to genetic resources and benefit sharing, and to the conservation and utilization of genetic resources
• strengthen multistakeholder arrangements for the effective implementation of farmers' rights and an ABS regime
• identify and strengthen institutional arrangements appropriate for farming communities to manage functions related to securing farmers' rights, enforcing an ABS regime and sustainable management of their genetic resources
• support innovative practices as a basis for implementing farmers' rights and ABS mechanisms, and for promoting conservation of biodiversity for livelihood security.

Their research consists broadly of two types of activities: community-level experimentation (through action research) and activities directly targeted at and involving relevant policymakers and other stakeholders. Figure 11.1 summarizes these activities.

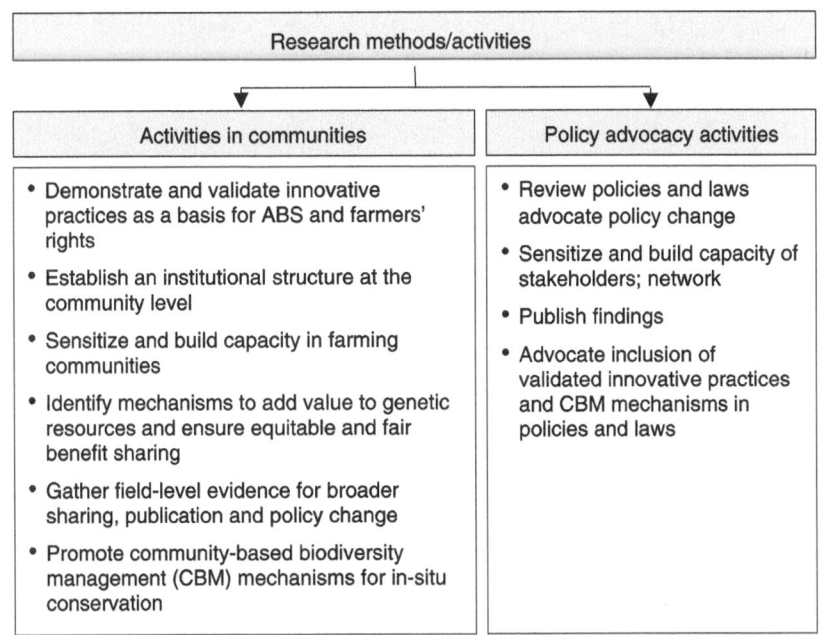

Figure 11.1 Action research framework

Demonstrating practices as a basis for implementing ABS and protecting innovative mechanisms

Through a process of learning by doing, the research team designed various mechanisms for in-situ conservation of biodiversity—the concept used to capture the range of activities geared toward improving agriculture through genetic resource management. These include both traditional mechanisms and novel ones. In practice, they are being tested "bundled together" in a strategy called community-based biodiversity management (CBM). CBM aims to demonstrate and validate basic ways to implement ABS and ensure farmers' rights. The mechanisms include practices useful for documentation, adding value, sustainable use and conservation, providing access and sharing the accrued benefits fairly and equitably (Figure 11.2). Another key area of activity is experimentation and research on novel institutional structures at the community and national levels that can carry out functions to ensure farmers' rights, including implementing a *sui generis* ABS regime.

At the heart of CBM is the idea that, in everyday farming practice, there are many entry points for ABS and they form an integral part of rural livelihoods. This represents a holistic and dynamic approach to ABS, rather than the narrow, overly legalistic approach that is so predominant in many ABS debates. It has taken many years of trial and error at the local level for the research team to put together such a strategy (Vernooy *et al.* 2009).

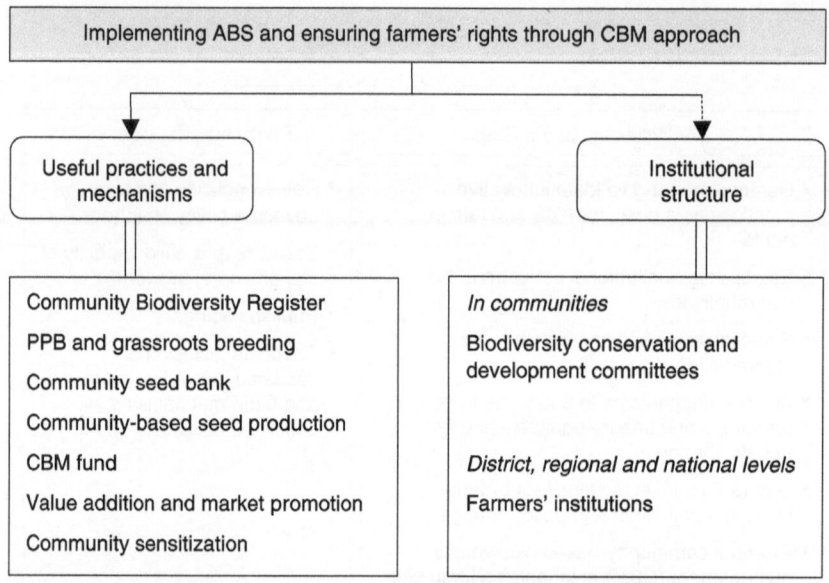

Figure 11.2 Community-based diversity management tools and approaches

Note: CBM = community-based management, PPB = participatory plant breeding.

Selection of research sites

A systematic approach to site selection was employed. Criteria and methods were proposed at a national-level stakeholder meeting and finalized by the research project steering committee. Criteria were developed for two levels: districts and village development committees (VDCs). Nepal has 75 districts, each with nine wards. VDCs are decentralized units within a district.

Criteria for district selection

- Choose districts where the Ministry of Forestry had piloted community biodiversity registration in 2003 (the plan was to reach 29 of the 75 districts)
- Cover as many types of genetic resources as possible, including high-yielding crops, non-timber forest products (NTFPs) and neglected and underused species (NUSs)
- Represent all ecological zones: terai (plains), hills and mountains
- Choose districts with communities that had experience in earlier work on biodiversity conservation.

Criteria for VDC selection

- Richness in biodiversity (major crops, NUSs and NTFPs)
- Dominance of indigenous and tribal groups in the population

Table 11.1 Selected research sites for demonstration and validation of practices

Genetic resources	Ecological zone	Districts selected	VDCs selected
Non-timber forest products	Mountains	Sangkhuwasabha	Tamaphok
Neglected and underused species	Hills	Dhading Kaski	Jogimara *Rupakot** *Lekhnath Municipality*
High production potential crops	Terai	Chitwan Bara	Bachhayauli *Kachorwa*

* Italics indicate VDCs with previous experience in in-situ conservation

- High dependency on natural resources, including genetic resources, for livelihood
- Interest in conservation and sustainable use of genetic resources
- Experience in the conservation and use of biodiversity
- Use and processing of and trade in globally important species.

The districts (Figure 11.3) were selected at a national stakeholder meeting. District-level stakeholder meetings were organized and four potential VDCs in each district were shortlisted, because of their experience working with various government bodies and NGOs. The research team visited all four and made an assessment using participatory rural appraisal tools, such as group discussions, focus group discussions and resource mapping, and Social Analysis System tools (SAS 2010), such as stakeholder identification and stakeholder analysis, to gather relevant information. Five research sites were then selected.

Establishment of biodiversity conservation and development committees

The establishment of a representative institution of farmers with a mandate to conserve and sustainably use genetic resources and associated traditional knowledge is a prerequisite for protecting the rights of communities during implementation of an ABS regime. Various institutional modalities were analyzed, but most local institutions were found to be member based and did not represent the whole community. Realizing the need for a system to represent the maximum possible number of households, a biodiversity conservation and development committee (BCDC) was established and supported as it became functional. The structure and organization of the pilot BCDC is shown in Figure 11.4. The structure is designed to support the local government in sustainably managing local biodiversity, including by implementing an ABS regime.

Awareness raising and capacity building

Raising the awareness of farming communities about the importance of biodiversity conservation and the scope of and policy issues related to ABS and

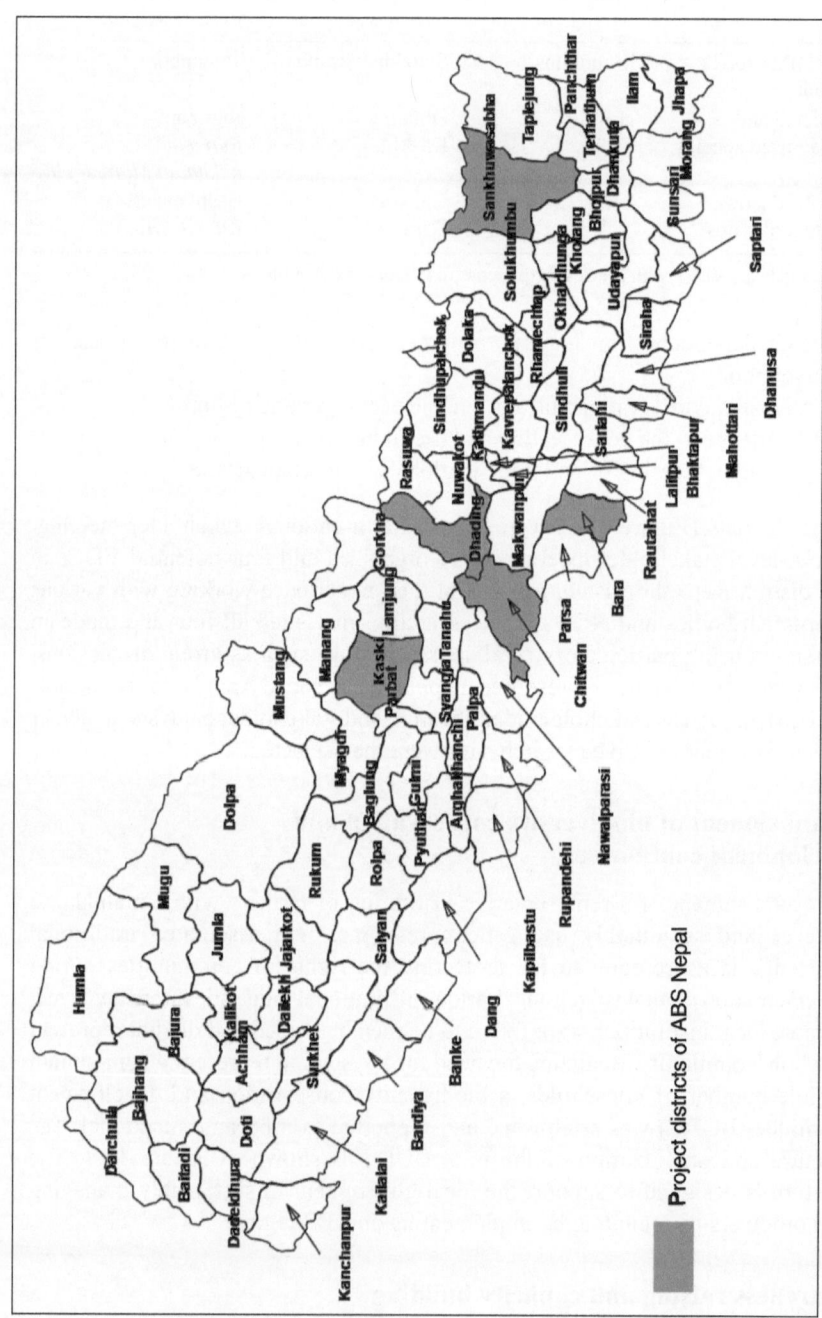

Figure 11.3 Location of the selected research districts

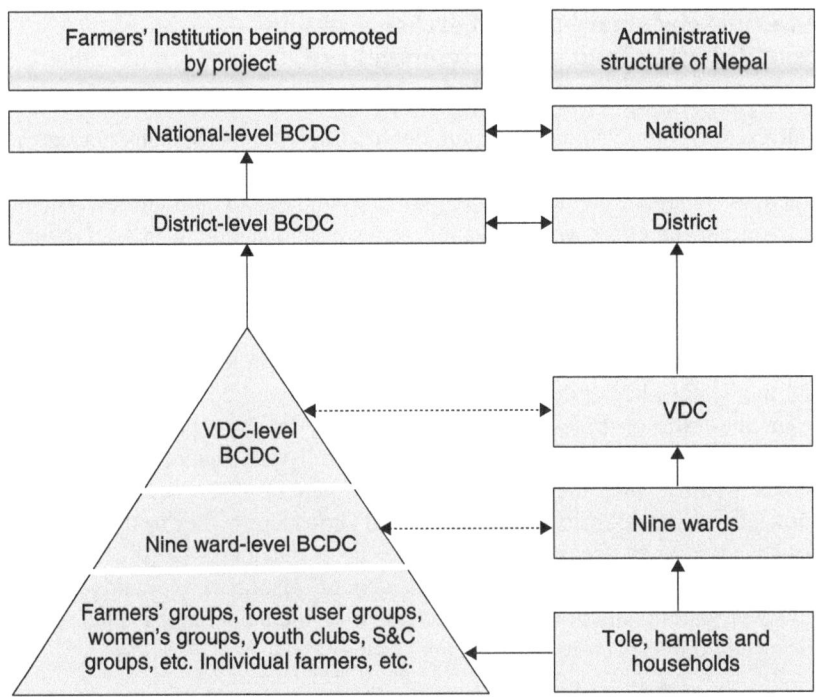

Figure 11.4 Structure of biodiversity conservation and development committees

Note: BCDC = biodiversity conservation and development committee, S&C = saving and credit, VDC = village development committee

farmers' rights was also part of the program. To do this, various types of events were organized and their effectiveness in disseminating a key message was evaluated. The events included village-level workshops, folk song competitions, biodiversity fairs, rural street dramas and farmers' level training in ABS and farmers' rights. The research team also focused on the preparation and distribution of farmer-friendly posters, briefing papers and policy briefs, which were found to be useful to literate people and students.

Activities were also supported to build the capacity of the BCDCs as community institutions that would implement an ABS regime and ensure farmers' rights. The team helped farmers understand their rights and responsibilities in biodiversity management, and supported development of the BCDCs through social mobilization and building capital. Members of the BCDCs were engaged in action research and exchange visits, and provided with opportunities to participate in national and international workshops, events and training. A community biodiversity management fund (Nepalese rupees (NPR) 500,000 for four BCDCs) was established as a revolving fund to share the benefits fairly and equitably and to make the institution sustainable.

Assessing customary use and exchange of and benefit sharing from genetic resources and indigenous traditional knowledge

The research team examined communities' traditional and customary management of genetic resources and indigenous traditional knowledge, as well as newer practices, adopted from development agencies and other communities. A number of traditional practices were relevant in terms of implementing an ABS regime at the study sites; most were related to the sustainable use of local genetic resources and associated traditional knowledge (Table 11.2).

Access to genetic materials is relatively free for the major cereal crops, NUSs and other materials of common use, but is restricted in terms of trade in NTFPs and medicinal plants (Table 11.3). Any benefits acquired from the use, added value and trade of these materials are generally taken as personal rather than collective income. Local communities are generally unaware of opportunities for benefit sharing when they provide access to genetic resources and traditional knowledge to outside researchers. Traditional knowledge owned by the community is viewed as open access knowledge—even to outsiders—and the traditional knowledge associated with most plant genetic resources is in this category. However, traditional knowledge associated with medicinal herbs is not as open, because of different customary practices (Table 11.4).

Mechanisms for ensuring ABS

Best practices at the community level, previously found to be useful in terms of conservation of local genetic resources on farm (Sthapit *et al.* 2008), have been adopted as a basis for ensuring farmers' rights and implementing an ABS regime in Nepal. Although some of these mechanisms have more than one function, we discuss them individually in the following sections, highlighting their principal role. Together, these eight mechanisms form a holistic and dynamic approach to ABS.

Table 11.2 Innovative and customary practices related to biodiversity conservation

Practice	Jogimara	Bachhayauli	Tamaphok
Seed saving, use, exchange and sale	✓	✓	✓
Community forest user group	✓	✓	✓
Leasehold forestry	✓		
Conservation blocks in forests	✓		
Mechanisms for sustainable harvesting			✓
Benefit-sharing mechanism			✓
Adding value to and marketing local plant genetic resources	✓	✓	✓
Community seed bank	✓		

Table 11.3 Access to genetic materials and extent of benefit sharing

Study VDCs	Source of genetic materials	In the community			Outside the community		
		Use/exchange (frequency)*	Value added/ trade (frequency)	Benefit sharing	Use/exchange (frequency)	Trade	Benefit sharing
Jogimara	NTFPs	NR (3)	NR (1)	Personal	Easy access (1)	No entry	–
	NUSs	NR (3)	NR (2)	Personal	Easy access (2)	NR	Personal
	Other crops	NR (3)	NR (1)	Personal	Easy access (2)	NR	Personal
Bachhayauli	NTFPs	NR (1)	NR (0)		Easy access (1)	NR	Personal
	NUSs	NR (1)	NR (0)		Easy access (1)	NR	Personal
	Other crops	NR (3)	NR (2)	Personal	Easy access (3)	NR	Personal
Tamaphok	NTFPs	NR (3)	NR (3)	Personal Community	No access (2)	No entry	–
	NUSs	NR (3)	NR (2)	Personal	Easy access (3)	NR	Personal
	Other crops	NR (2)	NR (1)	Personal	Easy access (2)	NR	Personal

Note: NR = not restricted, NTFPs = non-timber forest products, NUSs = neglected and underused species.
* 0 = never, 1 = rare, 2 = frequent, 3 = very frequent.

Table 11.4 Access to associated traditional knowledge and benefit sharing

Area of knowledge	Access	Use	Benefit	Customary practices
Medicinal herbs	Access to close relative, and successors	By the holders only	Personal benefit	Baidhya, Jhankri, Gurau*
Other plant genetic resources	Easy access	All	All	Free exchange

* The Baidhya, Jhankri and Gurau are the local people who have knowledge of the use of medicinal herbs for specific diseases.

Documentation of biodiversity and associated traditional knowledge

Communities were trained to document their genetic resources and associated traditional knowledge in a community biodiversity register (CBR). Sthapit *et al.* (2001) describe a CBR as "a record, kept in a register by community members, of the genetic resources in a community, including information on their custodians, passport data, agro-ecology, cultural and use values." A format for the CBR was developed, tested and approved by the Ministry of Forests and Soil Conservation (MoFSC) in consultation with LI-BIRD and the National Agriculture Research Council (NARC), and farmers were trained to use it. Elders and people with specific knowledge of genetic resources were much involved in the preparation of the CBR. If CBRs are recognized in policy and legal frameworks as certifying the custodians of genetic resources and associated traditional knowledge and they are compiled into a national CBR, this will facilitate bioprospecting, provide the basis for the ownership of genetic resources and associated traditional knowledge (Gauchan *et al.* 2005), and specify the community that must be involved in providing prior informed consent and in ABS.

Various institutional structures—community-based organizations, community forest user groups, cooperatives and BCDCs—were evaluated as maintainers of the CBRs (Subedi *et al.* 2005), and BCDCs were found to be the best accepted. Preparing a CBR at the ward level and compiling it at the VDC level was found to be the best procedure. The VDC CBRs will, in turn, be used to compile a national-level CBR. During the research project, 40 CBRs were prepared for four sites (36 at the ward level and four compiled at the VDC level); 45 were verified in total in the five VDCs (Table 11.5). Verification of CBRs was found to be essential. In some cases, this resulted in more varieties, while in other cases there were fewer.

Table 11.5 Species diversity documented in community biodiversity registers at the project sites

Crop category	No. of varieties or species before (and after) verification			
	Rupakot	*Tamaphok*	*Jogimara*	*Bachhayauli*
Cereals	80 (72)	104 (68)	104 (67)	NA (46)
Vegetables	61 (61)	91 (57)	145 (141)	NA (92)
Oil seeds	0 (0)	17 (11)	20 (20)	NA (8)
Pulses/legumes	40 (38)	10 (7)	63 (57)	NA (19)
Spices	0 (0)	17 (9)	34 (34)	NA (–)
Medicinal plants	143 (150)	155 (131)	157 (157)	NA (30)
Fruit	38 (38)	41 (38)	74 (72)	NA (31)
Other	29 (29)	24 (24)	39 (39)	NA (11)
Total	**391 (388)**	**459 (345)**	**636 (587)**	**NA (237)**

Note: NA = not available.

Empowering communities to make decisions about
conservation and use of genetic resources

Facilitated by the BCDCs and with support from the research team, communities were provided with the knowledge and skills they needed to analyze local biodiversity and prepare annual CBM plans. Four-cell analysis and other related tools were used to determine degree of biodiversity; then, village-level workshops are organized to finalize a CBM plan based on the biodiversity analysis. During the initial years, the project team provided technical and financial support. Earlier research had shown that the CBM approach is effective in empowering farming communities to apply a wide range of on-farm conservation practices (Subedi *et al.* 2006) that allow for maintenance or improvement of farmers' control over and access to crop genetic resources (Bragdon and Jarvis 2003). Thus, they ensure custodial rights.

Adding value to genetic resources and associated traditional knowledge

Marketing local genetic resources

Activities that add value and link farmers to markets provide an incentive for on-farm conservation of local genetic resources and traditional knowledge (Sapkota *et al.* 2006). Examples of such activities include the following.

- The project provided NRP 50,000 (about US$700) to the Tamaphok BCDC to purchase a share of the Tinjure Handmade Paper Industry (THPI), so that it will share in the regular profits of the paper industry and will use the proceeds for conservation and for the benefit of the community. The farmers' investment allowed THPI to expand the volume of its business. Thus, access to profitable shares may be an option for benefit sharing in a functional ABS framework.
- Support was given to local communities in Tamaphok to establish a "diversity block" of about 80 medicinal plant species. Now the communities have started to domesticate, add value to and market these herbs.
- The processing and marketing of local food products and Ayurvedic medicine prepared from local fruits and herbs, respectively, are being supported.
- A Lokta (*Daphne papyracea*) nursery was established in Tamaphok. Maintained by a women's group associated with the BCDC, it has produced so far about 15,000 saplings, about half of which are transplanted to open areas in the forest by nine community forest user groups.
- At the Jogimara and Begnash sites, seed production of local bean varieties is underway.

Incentives, either social or economic, are of crucial importance for the conservation of genetic resources in communities. As the addition of value and marketing of the genetic resources can be either individual or collective, the benefits from this activity can be distributed accordingly. Dividends from the THPI will be distributed to all the households in the VDC through the BCDC's programs, but the benefits of the other initiatives will remain with those who are involved in them. The people who have knowledge about growing, using, protecting and marketing the particular genetic resources are the ones who benefit. The benefits can be very location- and community-specific, as in the case of the Lokta nursery, as Lokta grows in a very specific habitat. Benefits can also be personal, as in the case of the processing of medicinal plants, where they go to the person who holds the knowledge of how to use the herbs.

Adding value to genetic resources by breeding

PPB and grassroots plant breeding have resulted in on-farm conservation of local rice varieties (Gyawali *et al.* 2006a,b), and many researchers see PPB as an essential step to securing the world's food supply (Vernooy 2003a). PPB also contributes to ensuring farmers' rights in various ways (Halewood *et al.* 2007) and promotes the sharing of benefits from the use of local genetic resources. Currently, Nepalese draft legislation accepts farmers as breeders of new varieties, but the process for confirming this is difficult and there is no incentive for farmers to continue this important work. Although varieties developed through PPB or grassroots breeding (e.g. Pokhareli Jethobudo, Barkhe 3004 and Sunaulo Suganda) have already been released by the National Seed Board, none is "owned" by any group. Thus, although farmers have been recognized as professional breeders—not a minor achievement—they receive no direct benefit from this work. Although there is a provision for claiming ownership, on paper, in the National Seed Act (1986), its scope and implications are not clear. Research has found that it is necessary to protect the varieties developed by farmers, but the contribution of other farmers who contribute to the genetic pool must not be undermined. Even the breeder farmers perceive that any incentive for them should not restrict the rights of other farmers to seeds. Above all, scientists and policymakers at all levels need to respect farmers as plant breeders in their own right (Vernooy 2003b).

Evidence from PPB and grassroots breeding was used to advocate suitable arrangements in a draft bill for the protection of plant varieties and farmers' rights, the seed act and seed regulations. Release of Pokhareli Jethobudo is an example of adding value to local genetic resources through selection of the best materials in an existing gene pool. In addition, some promising lines that are ready to release (Kachorwa 4 and Mansara 4) show how the value of local landraces can be increased through appropriate crossing with other varieties to improve or add certain traits that are favored by farmers.

Financing programs, such as PPB, may be a way to share non-monetary benefits from the use of genetic resources (Gauchan *et al.* 2006). The products of PPB were found to be readily available and everyone in the communities had benefited. For example, in Kachorwa, PPB varieties (K 4, K 5, K 125 and K 210) were found on about 12% of the area planted by households participating in PPB and these varieties were significantly more productive than their local parents.

Lines of Mansara and Biramphool crosses were grown on 8.5% of rice cultivation areas of the households who grow PPB varieties in Begnas; these varieties had traits favored by the farmers, but retained the original characteristics as well. This evidence shows that it is possible to conserve local landraces at the genetic level by adding value through breeding.

Community-based seed production

Community-based seed production (CBSP) of Jethobudo landraces was initiated with diverse stakeholders (Gautam 2008). About 29 farmers in the PPB group and other local farmers are also benefiting from production of seed from the PPB lines Mansara, Biramphool and Jhinuwa, and 13 farmers at the Kachorwa site are benefiting from seed production of the Kachorwa PPB lines. CBSP can be an incentive for breeder farmers as well as a source of benefits for other farmers who are only involved in seed production. PPB includes the downstream use and dissemination of products; thus, farmers may be involved not only in breeding activities, but also in the registration of varieties produced, their maintenance, seed multiplication and distribution and, as appropriate, commercialization. CBSP has been found to be the most successful mechanism for the use and dissemination of PPB products.

Benefits may also be generated from CBSP and marketing of promising local varieties. Currently, commercial production and marketing of nonregistered landraces is illegal, and the registration process for landraces is difficult; however, through CBSP, the production and exchange of seed of local landraces is being coordinated among farmers. CBSP may be an option for ensuring the rights of farmers to save, exchange and sell seed, reducing their dependency on the commercial seed supply and garnering respect for farmers as producers as well as consumers of seed.

Exchanging local genetic resources and knowledge

CBRs and community seed banks (CSBs) have also proved to be good community-level mechanisms for exchanging knowledge and genetic resources effectively (Mabille, n.d.; Shrestha *et al.* 2004) and have the potential to serve as a mechanism for sharing genetic resources nationally and internationally. CBRs are useful for locating the source of genetic material and identifying the holder of associated traditional knowledge (Subedi *et al.* 2006).

Farmers see the benefits of CSBs in terms of: getting local varieties of seed easily (70% of farmers); obtaining information about them (33%); benefiting from related CSB activities, such as access to the CBM fund (18%); conserving and obtaining varieties of rare seeds (12%); and other direct economic benefits (11%).

If CSBs could be linked with the national and international gene banks, they would be an effective model for exchanging genetic resources. Protecting community rights would be an important consideration. Combining CSBs with the CBRs and prior informed consent mechanisms would facilitate the exchange of genetic resources and associated traditional knowledge, while protecting the custodial rights of the community. Moreover, community-level events, such as biodiversity fairs, as piloted by LI-BIRD, would be very effective ways to exchange genetic resources and associated traditional knowledge within communities (Adhikari *et al.* 2004). Diversity fairs are indeed effective in creating awareness and interest among diverse stakeholders regarding the importance and value of local genetic resources. Such fairs were also found to be effective in communicating policy messages to a large number of people.

CSBs have been identified as a way to ensure farmers' rights and implement an ABS regime. Farmers see that CSBs have a positive effect in terms of access to local seed varieties; development of new varieties through PPB; identification, conservation and promotion of local landraces; protection of: the rights of communities to ownership of local genetic resources; strengthening of local and traditional seed systems; and promotion of self-storage of seed in households without affecting traditional seed exchange practices within and outside the villages (Table 11.6). Many of these functions strengthen the local seed system and, ultimately, ensure farmers' rights.

Table 11.6 Farmers' perception of the effect of community seed banks on the seed system*

	Negative (%)		No effect (%)		Positive (%)		Very positive (%)		Do not know (%)	
Access to seed of local varieties					59	(50)	37	(31)	23	(19.3)
Access to seed of improved varieties			18	(14.8)	66	(54.5)	10	(8.2)	27	(22.3)
Support for development of new varieties					66	(55.9)	19	(16.1)	33	(27.9)
Conservation of local landraces					65	(54.6)	23	(19.3)	31	(26)
Identification of local landraces					72	(61)	10	(8.4)	36	(30.5)
Protection of ownership of local genetic resources			2	(1.6)	64	(53.7)	5	(4.2)	48	(40.3)
Local seed system	1	(0.85)	11	(9.4)	54	(46.1)	2	(1.7)	49	(41.8)
Self-storage of the seed in household	14	(11.6)	29	(24.1)	45	(37.5)	2	(1.6)	30	(25)
Exchange of seed with neighbors	9	(7.62)	48	(40.6)	26	(22)	5	(4.2)	30	(25.4)
Exchange of seed with other villages	11	(8.66)	57	(44.8)	23	(18.1)	1	(0.78)	35	(27.5)

* 120 households: 90 involved in community seed banks, 30 not involved.

About 82% of farmers believe that a CSB can own the local landraces it maintains, and 69% also felt that a CSB can provide prior informed consent to outsiders on behalf of the farming communities.

Nepal's National Gene Bank is just becoming established. With good coordination, there is great potential to reduce the cost of collecting genetic resources from CSBs. Continuous development of genetic resources through in-situ conservation could also be ensured through proper linkage of CSBs with the National Gene Bank. Moreover, this would also give farmers easy access to a variety of genetic resources for domestication and repatriation (Majaju *et al.* 2003). The research team supported a CSB in Kachorwa to rejuvenate the seed of more than 80 local varieties of rice and the establishment of CSBs in Jogimara and Tamaphok.

Distributing the benefits

PPB and decentralized breeding, CSB and CBSP have all been shown to be effective ways to distribute benefits fairly and could form part of a formal benefit-sharing mechanism. In addition, monetary benefit sharing could be accomplished through a CBM fund, a revolving fund managed by BCDCs to enhance the livelihood of farmers and support conservation of biodiversity. Evidence shows that a CBM fund can be effectively operated as an incentive for farmers, especially minority groups and women, to be involved in conservation activities (UNDP/SGP 2009). A CBM fund is a means for communities to share benefits from common resources. Every household in a VDC is eligible for a loan from the fund. In the study area, funds are available in all wards and have benefited up to 90 households (7% of the total) in a year (Table 11.7). Money is distributed equitably to both men and women.

The project team is investigating whether monetary benefits from the use of community genetic resources can be deposited directly into the fund and used in biodiversity management and community welfare. National legislation is needed to allow this (in the ABS law), and the project team is advocating this.

It was also observed that establishment of a CBM fund builds the capacity of BCDCs in many ways. The funds increased the credibility of the institution, while also contributing to the livelihood of needy people. Most important, the CBM funds helped make BCDCs sustainable institutions in the community. The project team contributed NPR 0.45 million (about US$6,300) equally to three BCDCs (Jogimara, Bachhayauli and Tamaphok). BCDCs at the Rupakot and Kachorwa sites had already been provided with funds through previous LI-BIRD initiatives.

Communicating policy and legal messages to the community

Farmers and local communities perceive and interpret policies differently from decision-makers (Subedi *et al.* 2003). How farmers manage specific varieties on their farms is determined by many agro-ecological and socioeconomic factors,

Table 11.7 Use of the community biodiversity management fund at the project
sites in 2009

Site	Amount of fund (NPR)			No. of users of the fund			Coverage
	Project contribution	Community contribution	Total	Men	Women	Total	
Rupakot	–*						
Kachorwa	–*		450,000	10	80	90	All nine wards of VDC
Jogimara	150,000	15,300	165,300	29	21	50	All nine wards of VDC
Tamaphok	150,000	15,000	165,000	25	23	48	Members of 29 farmers' groups in all nine wards
Bachhayauli	150,000	15,000	165,000	In process of mobilization			

* Funds provided by other LI-BIRD projects prior to the current project.
Note: NPR 150,000 is about US$2,100.

including government policies that may have an impact. To be responsive to
different on-farm situations, the project team identified a variety of mechanisms
for communicating policy messages related to ABS.

Community-level events were organized to disseminate information about
biodiversity management and related policy messages: biodiversity fairs (four
events with over 200 participants and over 6,000 spectators), folk song competitions
(three events with about 5,500 spectators), rural street dramas (one event with nine
participants and more than 1,000 spectators), village-level workshops (four events
with 244 participants), joint monitoring visits and community-level training (six
events with 170 participant farmers).

Village-level workshops and training were found to be most effective for
communicating policy messages regarding biodiversity, but biodiversity fairs and
folk song competitions were useful in raising awareness among people of their
rights and responsibilities regarding conservation and use of genetic resources.
Various print materials, such as posters, brochures, comics and policy briefs, were
produced and distributed. A comparison of the various communication methods
used by the project team is presented in Table 11.8.

An effectiveness analysis of biodiversity fairs (Figure 11.5) showed that, on
average, respondents noticed about half the concepts (8.6 ± 0.5 out of 16) and
understood six (± 0.6) messages. The study showed that biodiversity fairs are an
effective tool for disseminating knowledge related to the concept and importance
of local biodiversity, for mobilizing people to conserve biodiversity and local

Table 11.8 Effectiveness of various communication methods based on users' judgment*

Method	Communicate policy message	Audience size	Interest of community	Participation	Effect	Cost
Folk song competition	2	5	5	5	2	3
Diversity fair	3	5	5	5	4	5
Rural street drama	3	3	4	2	3	3
Publication	4	2	2	1	2	4
Village-level workshops	4	2	3	3	3	3
Training	4	2	3	2	3	3

* Scores are on a scale of 1 to 5, where 1 = not at all effective and 5 = very effective.

resources, and in making people realize their rights to genetic resources and associated traditional knowledge.

Communicating research results

Sharing the results of the research has been crucial to developing another level of participatory action research and communicating the right messages in the policy dialog. Results were disseminated by facilitating the representation of the farmers in national and international forums, organizing joint monitoring visits and

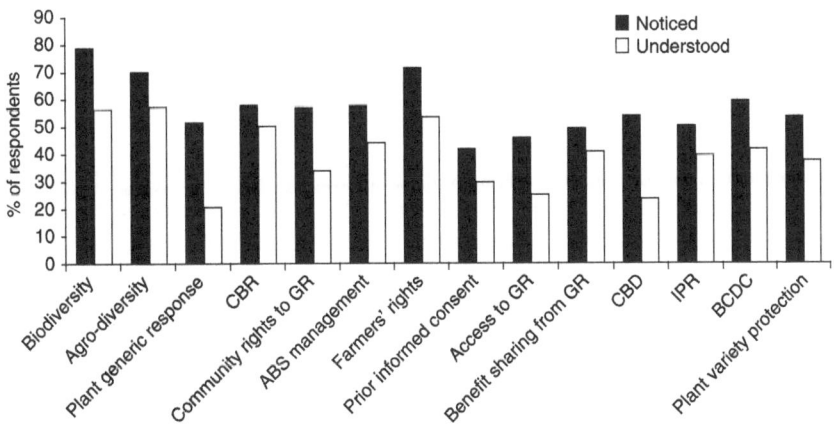

Figure 11.5 Effectiveness of biodiversity fairs in communicating messages (90 respondents)

Note: BCDC = biodiversity conservation and development committee, GR = genetic resources, CBD = convention of biological diversity, CBR = community biodiversity register, IPR = intellectual property rights

publications. Joint monitoring visits were found to be useful in policy advocacy efforts, as they gave Nepalese policymakers the chance to become more familiar with the everyday context of agricultural technology development and the problems farmers face. To share the information resulting from the research at the national level, various policy briefs, briefing papers and discussion papers were prepared and distributed. Moreover, the project team organized policy workshops, discussion forums and closed-group meetings with key policymakers to ask for appropriate changes in policies and legal frameworks based on their field evidence and reviews.

Review of constitutional provisions, policies and laws

As mentioned earlier, the concept of ABS is relatively new in Nepal. In February 1993, when the country became a party to the CBD, the government and NGOs began to discuss the importance of mainstreaming ABS issues in national policies. As a party to the CBD, Nepal was obliged to recognize and implement provisions to protect the rights of local and indigenous communities over their genetic resources. Soon after, following the developments taking place at the FAO's Commission on Plant Genetic Resources, national-level discussions were held on initiatives for the conservation, management and use of PGRFA and options to address farmers' concerns. Nepal also became a member of the WTO, which requires it to provide protection either by patents or by an effective *sui generis* system, or a combination thereof.

The Nepal Biodiversity Strategy (NBS) (Nepal 2002) paved the way for the MoFSC to draft the *Access to genetic resources and benefit sharing law* in 2002 and develop the Nepal Biodiversity Strategy Implementation Plan (2006–10). However, the law on ABS has not yet been implemented. A group of concerned stakeholders, including some indigenous NGOs, is discussing its provisions and making further recommendations for amendment. Similarly, the government has not been able to realize fully the goals of the NBS and the Nepal Biodiversity Strategy Implementation Plan.

With regard to plant variety protection and farmers' rights, building on the NBS and considering Nepal's membership in the WTO and its obligations under the agreement on Trade-Related Aspects of Intellectual Property Rights (TRIPS), the Ministry of Agriculture and Cooperatives (MoAC) has undertaken two major initiatives. The first is the enforcement of Nepal's Agricultural Biodiversity Policy (2006) and the second is the drafting of a law on the protection of plant varieties and farmers' rights, which, according to a 2005 WTO decision, the country must implement by June 2013.

More recently, after Nepal became a party to the International Treaty on Plant Genetic Resources for Food and Agriculture (ITPGRFA) on 10 January 2010, the MoFSC, together with NARC and the Seed Quality Control Centre (SQCC), strengthened collaboration with organizations, such as SAWTEE and LI-BIRD, to revise Nepal's Agricultural Biodiversity Policy (2007) and develop measures to implement the ITPGRFA at the national level. This is a welcome opening up of policy space.

In brief, effective policy and institutional mechanisms to ensure farmers' rights through supportive seed laws and regulations and implementation of an ABS regime are yet to be established in Nepal. Because of limited awareness of these issues (Gauchan *et al.* 2002), there is limited demand by stakeholders, including farmers. Moreover, there is a lack of well-accepted and verified approaches for realization of farmers' rights, including the sharing of the benefits arising from the use of genetic resources and associated traditional knowledge.

Interim Constitution of Nepal

Nepal's Interim Constitution (Nepal 2007) contains two major provisions that relate to ABS and the protection of the rights of local, indigenous and farming communities over their resources and associated traditional knowledge. First, in Part 3: Fundamental Rights, the Interim Constitution recognizes that "each citizen shall have the right to food sovereignty as provided for in the law" (Article 3.18(3)). As the "sovereignty framework" of food security is still a new and vague concept, a group of policymakers and stakeholders view the concept of the right to food sovereignty as not merely about guaranteeing food security, but also protecting farmers' rights over natural resources and biodiversity, including seeds.

Second, in Part 4: Responsibilities, Directive Principles and Policies of the State, the Constitution includes the following:

> 35 (18) The State shall pursue a policy of modernizing the traditional knowledge, skills and practices existing in the country by identifying and protecting them.
> 35 (5) ... the State shall also make arrangements for the special protection of the environment and the rare wildlife. Provision shall be made for the protection of the forest, vegetation and biodiversity, its sustainable use and for equitable distribution of the benefit derived from it.

Both of these provisions are crucial for the state to implement a pro-community ABS regime and devise mechanisms to protect the rights of local, indigenous and farming communities over natural resources, biodiversity and traditional knowledge. However, whether Nepal's ABS regime should be understood as the use of biological resources, in general, or deal only with genetic resources is still a major point of discussion at the national level.

Taking note of these issues and arguing that there is a need to strengthen government policy, a group of stakeholders, including lawyers and NGOs, has suggested the addition to the new Constitution of the following policies with regard to ABS:

- The state shall pursue a policy of conserving, managing and sustainably using genetic resources and traditional knowledge and of protecting the rights of local, indigenous and farming communities through an inclusive, fair and equitable ABS regime.

- The state shall promote a policy of ensuring appropriate and institutional representation and participation of local, indigenous and farming communities in policymaking and implementation processes on matters related to the conservation, management and sustainable use of genetic resources and traditional knowledge.

At present, constituent assembly members, political parties and various civil society groups are discussing constitutional measures to adequately protect the rights of local, indigenous, marginalized, deprived and poor communities and farmers. For example, SAWTEE and LI-BIRD have been proactively engaged in initiating local- and national-level discussions on ABS, traditional knowledge and genetic resources, so that a consensus on how the new Constitution should address these issues is developed for the collective welfare of the nation and its people, including local, indigenous and farming communities.

Nepal Biodiversity Strategy (NBS) and Implementation Plan

The NBS (Nepal 2002) was developed after consultations with a broad cross-section of stakeholders. It states that the Government of Nepal is committed to implementing the CBD at the national level and strongly dedicated to the protection and sustainable use of biologically diverse resources of the country, the protection of ecological processes and systems, and the equitable sharing of all benefits on a sustainable basis. The NBS underpins that "Long-term sustainable use of biological resources can only be achieved if the benefits are shared fairly and equitably, and the innovations, practices, and knowledge of indigenous peoples and local communities are respected" (Article 1.2.1(6)). It also recognizes the crucial role of PPB, gene banks and biodiversity registration in biodiversity conservation and sustainability, emphasizing the need to protect farmers' rights in light of the possible implications of intellectual property rights (IPRs). However, the government's strategies for developing and implementing proactive policies and laws on biodiversity conservation and an effective ABS regime are not clear in the NBS (see the box below). This important strategic document does not provide any clear-cut overview or guidance regarding the nature and scope of the ABS regime and the policy, legal and institutional measures needed to realize farmers' rights.

Institutional arrangements for coordination on biodiversity conservation

A 13-member National Biodiversity Coordination Committee (NBCC) was formed under the leadership of the minister of forests and soil conservation. Its members include representatives of concerned government ministries, the private sector, user groups, civil society, academic institutions and major donors. Five thematic subcommittees (forest, agriculture, sustainable use of biodiversity, genetic resources and biosecurity) have been created, and their coordinators are members of the NBCC.

> At the district level, district biodiversity coordination committees are being formed under the leadership of the chair of the district development committee. So far, only ten are in place, and the other 65 districts do not have a committee to implement local biodiversity strategies and action plans.

Source: MoFSC (2010).

In fact, for many issues surrounding IPRs and farmers' rights, the NBS conveys the message that there was no proper understanding of such issues when they were included in the document. For example, statements such as "Nepal will ensure the IPRs of farmers and local communities" and "farmers' rights will be vested in the international community as trustee for present and future generations of farmers and for supporting their continued contributions" do not provide any guidance in terms of implementation of IPRs and farmers' rights in the true spirit of national and international concerns. Even the Nepal Biodiversity Strategy Implementation Plan does not contain any clear action framework for establishing an ABS regime and the rights of local, indigenous and farming communities.

Legislative initiative for an ABS regime

Building on the NBS, the MoFSC prepared a draft law on access to genetic resources and benefit sharing in 2002. A series of consultations with concerned stakeholder groups, including indigenous community organizations at national and local level, was held, and input from these discussions was incorporated into the draft, which was submitted to parliament in 2006. However, the bill was not approved, as some parliamentarians, noting the reservations of a number of stakeholders, expressed the need to revise the proposed law. Addressing Nepal's obligations under the CBD, the bill deals with rules and regulations on access to genetic resources and sharing of the benefits arising out of the commercial use of genetic resources and traditional knowledge. It envisions the formation of a National Genetic Resources Conservation Council to be presided over by the secretary of the MoFSC as the national authority to implement an ABS regime in Nepal.

In line with the CBD, the bill states that the ABS mechanism has to be based on mutually agreed terms between the national authority and any third party. With regard to the sharing of derived benefits (from the third party) within Nepal, the bill provides for a tripartite monetary benefit-sharing mechanism, i.e. monetary benefits must be shared by the government, the national authority and the concerned local communities. According to the mechanism, if the government is the owner of a genetic resource that is commercialized by a third party, 50% of the benefits from its use accrue to the government, 30% to the authority and 20% to the communities. If a local community is the owner of the resource, 51% of the benefits go to the community, 29% to the authority and 20% to the government. The bill also provides that all parties are required to share 10% of their benefits with the concerned local government institution. To ensure that community members do not face problems or engage in possible conflicts over the benefits they receive, the bill states that the sharing of the benefits among the communities will be done through the concerned local government institution.

The bill also specifies certain prior informed consent conditions with regard to the documentation of and access to genetic resources and traditional knowledge. Prior informed consent must be obtained from the local communities before documentation or registration of their resources and traditional knowledge. The bill mentions that the government must publish the third party's proposal in national and local newspapers, and the concerned village development committee or municipality must conduct a public hearing in the concerned communities so that prior informed consent can be assured. In addition, the bill requires formation of a negotiating committee, made up of members of all concerned stakeholders, including local communities, to discuss the third party's proposal on access to genetic resources and traditional knowledge.

In recent years, a number of organizations, mainly those of indigenous communities, have been demanding a substantial revision of the ABS bill so that their concerns with regard to the protection of specific indigenous rights are addressed. Organizations such as SAWTEE, LI-BIRD, Forum for Protection of Public Interest, USC Canada Asia, ActionAid International Nepal and a number of lawyers' associations are helping the government and community groups identify adequate strategies and measures so that the bill addresses the relevant concerns and there is no further delay in the implementation of Nepal's ABS regime.

Agricultural Biodiversity Policy

Taking note of the significance of a policy for promoting conservation and development of agricultural genetic resources at ex-situ and in-situ conservation levels, the NBS highlights several agricultural biodiversity issues, including those concerned with ABS, traditional knowledge and farmers' rights. Building on the NBS and the National Agriculture Policy (2004), the MoAC prepared Nepal's Agricultural Biodiversity Policy (2007). It seeks to "protect the rights and welfare of farming communities" (Article 4(a)), including their "indigenous knowledge, skills and technologies" (Article 4(b)). It also aims to "develop options for a fair and equitable sharing of benefits arising from the access and use of agricultural genetic resources and materials" (Article 4(c)).

However, the policy was developed only to meet the country's obligations under the CBD and the NBS. There is growing realization by the government and concerned stakeholders, including national and local NGOs, that it needs extensive revision. Although the policy's current objectives are to give special priority to the realization of farmers' rights over agricultural genetic resources and traditional knowledge, many of its strategies are not harmonized and in some cases are even contradictory: for example, with regard to the issue of the ownership over agricultural genetic resources.

Thus, as suggested and mandated by a meeting of the National Agricultural Biodiversity Conservation Committee, formed to ensure the policy's implementation, a Technical Committee was created in February 2010 to revise the policy. This committee includes members from the MoAC, NARC, SQCC, SAWTEE, LI-BIRD and USC Canada Asia. At its first meeting in March 2010, noting recent global and national developments on ABS and farmers' rights issues, its members agreed to incorporate new policy strategies, for example, with regard to the conservation and development of PGFRA at in-situ and ex-situ levels and the protection of farmers' rights over PGFRA and traditional knowledge. As Nepal became a party to the ITPGRFA in January 2010, the committee also considered this treaty's provisions and Nepal's obligations.

Plant variety protection and seed laws

Nepal does not have a specific policy or law addressing plant variety protection or breeders' rights. Seed development, certification, registration and release are administered under the Seed Act and Regulations (1988) and subsequent amendments. To qualify for registration and release, seeds must be distinct from other varieties, uniform and stable. In addition, the National Seed Policy (1999) focuses on variety development, maintenance, seed supply and private-sector participation in seed commerce and quality control.

The National Seed Board, created under the Seed Act, is responsible for regulating and controlling the quality of seeds produced by private and government seed companies. In addition, subject to specific terms and conditions, the board has the authority to: approve, release and register seeds; assess their distinctness, uniformity and stability; and provide ownership certificates to the breeders (the scope of such ownership is, however, not clear and this is a voluntary requirement). Generally, modern varieties of seed coming from research facilities to farmer fields are supported by certification. Seeds produced by local farmers do not qualify for certification, although these are usually farmer-preferred varieties, which may include both landraces and modern varieties. There is growing demand by local-level organizations and groups for flexibility of the existing seed legislation and for the government to develop support and promotional measures—for example, through participatory variety selection and PPB programs—to enable farmers to benefit from seed development and trade.

During negotiations surrounding WTO membership, as part of its obligation to comply with the TRIPS agreement, the Government of Nepal made a commitment to devise a new and separate law on plant variety protection and to amend existing seed legislation to harmonize it with new IPR rules. In recent years, many agencies, including seed companies in Nepal and elsewhere, have been pressing the government to enact a plant variety protection law based on the UPOV model. Some have also been calling on Nepal to join UPOV, arguing that it would attract multinational seed companies to invest in plant breeding and seed development in the country. However, the government, policy and research experts from NARC and a number of active civil society organizations, such as ActionAid Nepal,

SAWTEE, LI-BIRD and the Forum for Protection of Public Interest, and farmers' groups at the local level, are convinced that there is no need to join UPOV, and Nepal would do well to enact a *sui generis* act to protect plant varieties and farmers' rights.

So far, the MoAC, in consultation with relevant stakeholder groups, including private seed companies, NARC, SQCC and civil society groups, has drafted a law on plant variety protection and farmers' rights. It is encouraging that the government has expressed a willingness to work with civil society organizations to develop such legislation. Government and these groups are now working on measures to protect farmers' rights over IPR-protected varieties and to ensure that farmers are able to register their varieties and related knowledge in the formal system and obtain benefits. The MoAC and SQCC, in collaboration with SAWTEE and LI-BIRD, have also initiated an informed multi-stakeholder dialog, so that there is proactive consensus on the scope, nature, objectives and other provisions of the plant variety protection and seed laws.

Conclusions

In Nepal, the major obstacles to achieving a fair and equitable ABS regime and protecting the rights of local, indigenous farming communities are:

- limited documentation and registration of genetic resources and traditional knowledge and limited scientific knowledge about such documentation and registration at national and local levels
- lack of information about the use and value of genetic resources and traditional knowledge and limited focus on community-based biodiversity management projects and participatory programs, such as participatory variety selection and PPB
- limited national capacity to regulate or promote bioprospecting as well as biotechnology
- lack of awareness about IPRs, biotechnology and community rights over genetic resources and traditional knowledge
- limited institutional arrangements for facilitating access, prior informed consent and benefit sharing at both government and community levels; for example, a vision for the effective operationalization of community and national gene funds has yet to appear and mature at the policy and institutional level, the national-level biodiversity conservation committee has yet to become active and local-level committees have yet to be formed and institutionalized across the country.

To realize farmers' rights over PGRFA and traditional knowledge, including implementing an ABS regime, adequate and effective arrangements at three levels—policy, institutional and practice—are a must. The major constraint is that the government has yet to build and implement the policy, institutional and practical mechanisms to balance the rights of breeders and farmers.

The role that participatory variety selection and PPB can play in the development of varieties as well as the realization of farmers' rights over seeds has yet to be recognized at the policy level. The strategic directions in this regard will have to take national and local issues into consideration; however, government efforts have been very limited, mainly because of a lack of awareness among policymakers and resources. Moreover, the private sector remains unconvinced that such policies and laws will not favor farmers' rights at their expense.

Enhancement of the capacity of government, as well as community institutions, to register genetic resources and traditional knowledge, to ensure prior informed consent, to distribute, use and mobilize benefit-sharing funds, and to negotiate mutually agreed terms with third parties is also critical. Some progress is being made, but there is still a long road to walk.

References

Adhikari, A., Rana, R.B., Sthapit, B., Subedi, A., Shrestha, P.K., Upadhaya, M.P., Baral, K.P., Rijal, D. and Gyawali, S. (2004) Effectiveness of diversity fair in raising awareness of agrobiodiversity management. In B.R. Sthapit, M.P. Upadhaya, P.K. Shrestha and D.I. Jarvis (eds.), *On-farm conservation of agricultural biodiversity in Nepal. Volume II, Managing diversity and promoting its benefits*. Proceedings of the second national workshop, 25–27 August 2004, Nagarkot, Nepal. Local Initiatives for Biodiversity, Research and Development, Pokhara, Nepal. pp. 236–45. Available at: http://idl-bnc.idrc.ca/dspace/bitstream/10625/36097/1/122787_v2.pdf (accessed 29 March 2011).

Bragdon, S. and Jarvis, D.I. (2003) The role of legislature, policy and agrobiodiversity management on-farm. In D. Gauchan, B.R. Sthapit and D.I. Jarvis (eds.), *Agrobiodiversity conservation on-farm: Nepal's contribution to scientific basis for national policy recommendation*. International Plant Genetic Resources Institute, Rome, Italy. p. 2. Available at: www.bioversityinternational.org/fileadmin/bioversity/publications/pdfs/864.pdf?cache=1301496978 (accessed 30 March 2011).

CBS (Central Bureau of Statistics) (2003) Population census survey of Nepal. CBS, Kathmandu, Nepal.

Gauchan, D., Adhikari, A., Shrestha, P.K. and Sthapit, B. (2006) Impact of market and value addition initiatives on crop diversity and livelihood improvement in western mid-hills of Nepal. In B. Sthapit and D. Gauchan (eds). *On farm management of agricultural biodiversity in Nepal: lesson learned*. Proceedings of a national symposium, 18–19 July 2006, Kathmandu, Nepal. Local Initiatives for Biodiversity, Research and Development, Pokhara, Nepal. pp. 196–206.

Gauchan, D., Baniya, B.K., Upadhyay, M.P. and Subedi, A. (2002) National plant genetic resource policy for food and agriculture: a case study of Nepal. International Plant Genetic Resources Institute, Rome, Italy.

Gauchan, D., Sthapit, B. and Subedi, A. (2005) Community biodiversity register: a review of South-Asian experience. In A. Subedi, B. Sthapit, M. Upadhaya and D. Gauchan (eds.), *Learnings from community biodiversity register in Nepal: proceedings of national workshop, 27–28 October 2005*. Local Initiatives for Biodiversity, Research and

Development, Pokhara, Nepal. pp. 3–13. Available at: http://idl-bnc.idrc.ca/dspace/bitstream/10625/27838/1/122788.pdf (accessed 29 March 2011).

Gautam, J.C. (2008) *Country report on the state of the Nepal's plant genetic resources for food and agriculture.* Commission on Genetic Resources for Food and Agriculture, Food and Agricultural Organization, Rome, Italy. Available at: ftp://ftp.fao.org/ag/agp/country reports/NepalFinalReport.pdf (accessed 16 March 2011).

Gyawali, S., Sthapit, B., Bhandari, B., Bajracharya, J., Shrestha, P.K., Upadhaya, M. and Jarvis, D. (2006a) Jethobudo landrace enhancement for in-situ conservation of rice in Nepal. In B. Sthapit and D. Gauchan (eds.) *On farm management of agricultural biodiversity in Nepal: lesson learned.* Proceedings of a national symposium, 18–19 July 2006, Kathmandu, Nepal. Local Initiatives for Biodiversity, Research and Development, Pokhara, Nepal. pp. 146–63.

Gyawali, S., Sthapit, B., Bhandari, B., Shrestha, P., Joshi, B.K., Mudwori, A., Bajracharya, J. and Shrestha, P. (2006b) Participatory plant breeding (PPB): a strategy for on-farm conservation and improvement of landraces. In B. Sthapit and D. Gauchan (eds.), *On farm management of agricultural biodiversity in Nepal: lesson learned.* Proceedings of a national symposium, 18–19 July 2006, Kathmandu, Nepal. Local Initiatives for Biodiversity, Research and Development, Pokhara, Nepal. pp. 164–73.

Halewood, M., Deupmann, P., Sthapit, B.R., Vernooy, R. and Ceccarelli, S. (2007) *Participatory plant breeding to promote farmers' rights.* Bioversity International, Rome, Italy.

Mabille Y. (ed.) (n.d.) Farmers as banker: community seed banks (issue paper). Deutsche Gesellschaft für Technische Zusammenarbeit, Eschborn, Germany. Available at: www.agrobiodiversity.cn/uploads/media/Farmers_as_bankers-Community_seed_banks.pdf (accessed 29 March 2011).

Majaju, C., Zinhanga, F. and Rusike, E. (2003) *Community seed banks for semi-arid agriculture in Zimbabwe.* Centro Internacional de la Papa—Users' Perspectives With Agricultural Research and Development, Manila, Philippines. pp. 294–301. Available at: www.eseap.cipotato.org/upward/Publications/Agrobiodiversity/pages%20294-301%20 (Paper%2038).pdf (accessed 29 March 2011).

MOF (Ministry of Finance) (2008) Economic survey, fiscal year 2007/08. MOF, Kathmandu, Nepal.

MoFSC (Ministry of Forests and Soil Conservation) (2002) *National biodiversity strategies.* MoFSC, Kathmandu, Nepal.

—— (2007) Website of the MoFSC. Government of Nepal, Kathmandu, Nepal. Available at: www.mofsc.gov.np (accessed 10 May 2010).

Nepal, Government of (2002) *Nepal biodiversity strategy.* Government of Nepal, Kathmandu, Nepal. Available at: www.ansab.org/UserFiles/Nepal%20Biodiversity%20 Strategy%202002.pdf (accessed 28 March 2011).

—— (2007) Interim constitution of Nepal. Government of Nepal, Kathmandu, Nepal. Available at: www.worldstatesmen.org/Nepal_Interim_Constitution2007.pdf (accessed 28 March 2011).

Sapkota, T., Bhandari, B., Regmi, B., Rijal, D., Subedi, A., Shrestha, P., Gauchan, D., Poudel, I., Subedi, S., Tamang, B.B. and Tiwari, R. (2006) Marketing of agro-biodiversity in Nepal: option for on-farm conservation of agro-biodiversity through market incentives. In B. Sthapit and D. Gauchan (eds.), *On farm management of agricultural biodiversity in Nepal: lesson learned.* Proceedings of a national symposium, 18–19 July 2006, Kathmandu, Nepal. Local Initiatives for Biodiversity, Research and Development, Pokhara, Nepal. pp. 196–205.

SAS (Social Analysis Systems) (2010) Social Analysis Systems. SAS, Ottawa, Canada. Available at: www.sas2.net (accessed 17 March 2011).

Shrestha, P., Subedi, A., Rijal, D., Singh, D., Sthapit, B.R. and Upadhaya, M.P. (2004) Enhancing local seed security and on-farm conservation through community seed bank in Bara district of Nepal. In B.R. Sthapit, M.P. Upadhaya, P.K. Shrestha and D.I. Jarvis (eds.), *On-farm conservation of agricultural biodiversity in Nepal. Volume II, Managing diversity and promoting its benefits*. Proceedings of the second national workshop, 25–27 August 2004, Nagarkot, Nepal. Local Initiatives for Biodiversity, Research and Development, Pokhara, Nepal. pp. 70–6. Available at: http://idl-bnc.idrc.ca/dspace/bitstream/10625/36097/1/122787_v2.pdf (accessed 29 March 2011).

SGP (Small Grants Programme) (2009) *Strengthening the capacity of community seed bank for enhancing local seed security and agrobiodiversity conservation in Central Terai*. Global Environment Facility, Washington, DC, USA. Available at: http://sgp.undp.org/web/projects/9740/strengthening_the_capacity_of_community_seed_bank_for_enhancing_local_seed_security_and_agrobiodiver.html (accessed 29 March 2011).

Sthapit, B.R., Joshi, K.D., Rana, R.B., Upadhayay, M.P., Eyzaguirre, P. and Jarvis, D. (2001) *Enhancing biodiversity and production through participatory plant breeding. An exchange of experience from South and South East Asia*. Proceedings of the international symposium on PPB and PGR enhancement. PRGA Program, Cali, Colombia. pp. 29–54.

Sthapit, B., Gauchan, D., Subedi, A. and Jarvis, D. (eds.) (2008) *On farm management of agricultural biodiversity in Nepal: lesson learned*. Proceedings of a national symposium, 18–19 July 2006, Kathmandu, Nepal. Local Initiatives for Biodiversity, Research and Development, Pokhara, Nepal.

Subedi, A., Gauchan, D. and Sthapit, B. (2003) Developing policies on agricultural biodiversity conservation and use in Nepal. In *Conservation and sustainable use of agricultural biodiversity: a sourcebook*. Centro Internacional de la Papa—Users' Perspectives With Agricultural Research and Development, Manila, Philippines. pp. 553–8. Available at: www.eseap.cipotato.org/upward/Publications/Agrobiodiversity/pages%20553-558%20(Paper%2066).pdf (accessed 29 March 2011).

Subedi, A., Shrestha, P., Shrestha, P., Gautam, R., Upadhaya, M., Rana, R.B., Eyzaguirre, P. and Sthapit, B. (2006) Community biodiversity management: empowering community to manage and mobilize agricultural biodiversity. In B. Sthapit and D. Gauchan (eds.), *On farm management of agricultural biodiversity in Nepal: lesson learned*. Proceedings of a national symposium, 18–19 July 2006, Kathmandu, Nepal. Local Initiatives for Biodiversity, Research and Development, Pokhara, Nepal. pp. 140–5.

Subedi, A., Sthapit, B., Shrestha, P., Gauchan, D. and Upadhaya, M. (2005) Emerging methodology of community biodiversity register: a synthesis. In A. Subedi, B. Sthapit, M. Upadhaya and D. Gauchan (eds.), *Learnings from community biodiversity register in Nepal: proceedings of national workshop, 27–28 October 2005*. Local Initiatives for Biodiversity, Research and Development, Pokhara, Nepal. pp. 75–83. Available at: http://idl-bnc.idrc.ca/dspace/bitstream/10625/27838/1/122788.pdf (accessed 29 March 2011).

Vernooy, R. (2003a) *Seeds that give: participatory plant breeding*. International Development Research Centre, Ottawa, Canada. Available at: http://idl-bnc.idrc.ca/dspace/bitstream/10625/25864/8/118482.pdf

—— (2003b) Food security: seeds of threat, seeds of solutions. International Development Research Centre, Ottawa, Canada. Available at: http://idl-bnc.idrc.ca/dspace/bitstream/10625/28903/1/118910.pdf

Vernooy, R., Shrestha, P., Ceccarelli, S., Ríos, H., Song, Y. and Humphries, S. (2009) Towards new roles, responsibilities and rules: the case of participatory plant breeding. In S. Ceccarelli, E. Guimaraes and E. Weltzien-Rattunde (eds.), *Plant breeding and farmer participation.* Food and Agriculture Organization of the United Nations, Rome, Italy. pp. 613–28.

12 Conclusions

Race to the bottom versus slow walk to the top

Ronnie Vernooy and Manuel Ruiz

Making local experience count

While, for almost a decade at international and national fora, decision-makers, advisors and advocates have ardently debated the precise wording of ABS policies and laws, in communities around the world, farmers, indigenous peoples, researchers, NGO staff and local government agents have been working hard to implement, test and assess effective, fair and equitable mechanisms. Often, but not always, they established and tried to maintain direct links with the national and international fora to have, at least, a voice in the debates and negotiations and, in the best of cases, a choice as well. In this chapter we review the experiences and lessons learned from the case studies in light of the broader policy and legal processes, summarized in Part 1. We do this by re-examining the questions we set out to answer in Part 1 from a comparative perspective. This will highlight the relevance of the cases to both the Convention on Biological Diversity (CBD) and the International Treaty on Plant Genetic Resources for Food and Agriculture (ITPGRFA).

In the last decade, without doubt, ABS issues have increasingly become part of the international and national policy and legal agendas related to biodiversity in general and genetic resources in particular. As such, fair and equitable ABS has acquired what could be called a formalized nature. Some have called this process "the race to the bottom." However, at the local level, many of the issues continue to be dealt with through customary (sometimes formalized, often informal) practices, rules and regulations. Benefit sharing has been taking place *regardless* of policy and legal mandates, and sometimes in *opposition* to ineffective or unfair policies and laws.

While the recent adoption of the Nagoya Protocol (COP 2010) and the ITPGRFA (FAO 2009) signal important progress, much remains to be done to make ABS work in practice. This is particularly so in the case of genetic resources in agriculture. The cases in this book offer a number of suggestions derived from learning by doing, making mistakes and trying again.

Key features of the cases

We set out to find answers to five interrelated questions that are at the heart of ABS. We present them again here and provide short but succinct answers. Before doing that, we first present a summary of the key ABS features of the case studies. These features begin with recognition—the ways in which the knowledge and experience of the custodians of biodiversity are being recognized, assessed and valued. Proper recognition is a precondition for any meaningful ABS mechanism or regime. The case studies make this very clear. In policy and legal spheres, using this principle as a basis remains a work in progress. Some countries, such as Honduras among the cases presented here, continue to disregard it, while others, such as Peru, Nepal and China, are adopting a more positive attitude. Access—as a precondition for fair benefit sharing—then benefit sharing follow, and supportive policies and laws complete the framework. As Table 12.1 illustrates, there are a number of similarities, but also important differences among the cases.

Similarities can be found in the principles that inform the efforts, e.g. farmers and indigenous people are knowledgeable and have much to contribute to conservation and improvement of genetic resources, but can also greatly benefit from interaction and cooperation with others (all cases), as well as in the entry points for action, e.g. addressing biopiracy (Peru), introducing and expanding PPB (all cases except Peru), promoting in-situ conservation of genetic resources (all cases) and attempts to open direct dialog with key policy- and lawmakers (all cases except Syria and Honduras).

Differences concern the scope of the efforts, e.g. from a focus on biopiracy (Peru) or PPB (Syria) to a broader perspective on local agricultural development (Cuba, Nepal) and the policy and legal context in which the actions are taking place—ranging from constraining or even hostile (Honduras) to more open (Jordan) and encouraging and receptive (Nepal, China), and from little or no space for policy dialog (Syria, Honduras) to direct, collaborative policy experimentation (China).

Answering the questions

Local perceptions/national definitions

How do people at the local level perceive and assess ABS questions, especially in light of national and international guidelines, model laws and other new forms of defining and regulating ABS regarding biodiversity resources?

ABS issues have only quite recently reached local communities, sometimes through negative experiences, such as biopiracy, and sometimes through positive ones, such as the participatory action and development efforts to conserve agricultural biodiversity and improve genetic resources highlighted in this book. To a considerable extent, farmers' local-level agricultural practices (i.e. seed exchange, community fund) have traditionally been based on some form of equitable benefit sharing that stems from regular and ancestral practices in some cases, rather than from specific legal mandates or ABS talk. It is only since the establishment of the CBD that a legal obligation has arisen in this regard.

Table 12.1 Comparing the cases: ABS in practice

	Recognition	Access	Benefit sharing	Policies and laws*
Peru	Legal recognition of traditional knowledge Legal recognition of biodiversity registers	Strengthened capacities of national agencies that govern ABS policies and laws	Communities and their seeds are recognized as key actors in conservation efforts Gastronomy and local/native inputs are key drivers of and demand for revival of and demand for native seeds and produce	Implementation of existing framework on ABS and protection of traditional knowledge National Commission for the Prevention of Biopiracy created by law Party to the CBD since 1993 Party to the ITPGRFA since 2003
Syria	Farmers' traditional knowledge and practices regarding genetic resources	Decentralized plant breeding	Capacity building among farmers Women involved through special activities Improved varieties Increased crop diversity Improved seed system Seed production and commercialization (nascent)	Ministerial decree (1975) regulating variety release and seed multiplication and distribution Seed Law being drafted Party to the CBD since 1996 Party to the ITPGRFA since 2003
Jordan	Farmers' traditional knowledge and practices regarding genetic resources Awareness of roles of farmers as breeders	Decentralized plant breeding	Capacity building among farmers Improved varieties Seed production and commercialization (nascent)	Law for the protection of new plant varieties since 2000 Agriculture Law No. 44 (2002) provides the framework for variety release and registration, seed production, quality control and seed trade Member of UPOV since 2004 Party to the CBD since 1993 Party to the ITPGRFA since 2002

(Continued)

Table 12.1 (Continued)

	Recognition	Access	Benefit sharing	Policies and laws*
Honduras	Farmers' traditional knowledge and practices regarding genetic resources Farmers as bona fide seed producers Local agricultural research committees (CIALs) Release of participatory plant varieties at the municipal level	PPB (links to improved materials) Exchange of breeding materials among CIALs and communities Association of CIALs as a platform for exchange of breeding materials	Capacity building among farmers Improved varieties Seed production and commercialization (nascent)	Law for protection of plant varieties under review since 2001 National Committee on Biosafety; biosafety law not yet approved Signatory to Cartagena Biosafety Protocol Party to the CBD since 1995 Party to the ITPGRFA since 2004
China	Farmers' traditional knowledge and practices regarding genetic resources (especially women farmers) Community biodiversity registers (CBRs) Farmers as bona fide seed producers	PPB (links to improved materials) Diversity fairs Community seed banks	Capacity building among farmers Improved varieties Community seed banks Seed production and commercialization (nascent) Within existing legislative system, farmers and breeders reach formal agreement on distribution of monetary and non-monetary benefits facilitated by a third party	High-level advocacy Informing key policies and laws Member of UPOV since 1999 Regulation on Plant New Variety Protection (1997) Seed Law (2000) Rules for Management of Genetic Resources of Agricultural Crops (2003) Outline of National Biological Species Resources Conservation and Utilization Plan (2007) Science and Technology Progress Law (2008) National Bio-diversity Protection Strategy and Action Plan (2009) Revised Patent Law (2009) Party to the CBD since 1993 Not a party to the ITPGRFA

Cuba	Farmers' traditional knowledge and practices regarding genetic resources Farmers as bona fide seed producers Local groups of farmer experimenters and seed producers	Participatory variety selection Participatory seed diffusion Diversity seed fairs Community seed banks	Improved varieties Increased crop diversity Seed production and commercialization	Constitution (1976) Resolution no. 111/96 (1996) Environment law (1997) Biosafety law (1999) Farmer-improved varieties formally released Party to the CBD since 1994 Party to the ITPGRFA since 2004
Nepal	Farmers' traditional knowledge and practices regarding genetic resources Farmers as bona fide seed producers Awareness of importance of biodiversity, e.g. folk song competition, street drama CBRs Community-based biodiversity conservation and development committees (BCDCs) and plans Farmers' rights as effective political and legal concept	Strengthened new forms of organizations, e.g. BCDCs, and institutions, such as BCDC plans Community seed banks PPB (links to improved materials)	Capacity building among farmers Improved varieties Community seed banks BCDCs and plans Domestication of wild species Marketing Seed production and commercialization Nurseries Community biodiversity management funds	High-level advocacy Informing key policies and laws CBRs recognized in policy/law National gene bank under development Farmers accepted as bona fide breeders PPB varieties formally released Seed Act being updated Protection of Plant Varieties and Farmers' Rights law (*sui generis*) in preparation Nepal Agricultural Biodiversity Policy under development Nepal Biodiversity Strategy Implementation plan in progress Draft ABS law Party to the CBD since 1993 Party to the ITPGRFA since 2009

Note: BCDC = Biodiversity conservation and development committee, CBD = Convention on Biological Diversity, CBR = community biodiversity register, CIAL = local agricultural research committee, ITPGRFA = International Treaty on Plant Genetic Resources for Food and Agriculture, PPB = participatory plant breeding, UPOV = Union for the Protection of New Varieties of Plants

* For information concerning parties to the CBD, see CBD (n.d.) and for parties to the ITPGRFA, see FAO (n.d.).

Concerns about ABS have often appeared at the crossroads of two forces—emerging from local practical experience and insights on one hand (e.g. after 5–10 years of PPB efforts when new varieties were developed and teams were wondering what to do with them), and awakened by developments in national and international arenas on the other (countries actively seeking to develop ABS policies, e.g. Peru, and, more recently, Nepal and China). As exemplified by the case of Jordan, the evolution of PPB led almost naturally to the realization by the research team that breeding programs are not just a matter of technical expertise, but also that important policy and legal aspects have an impact on PPB. In the current policy context, these aspects are being phrased in terms of ABS questions, including such fundamental questions as who owns or has property rights over seeds and breeding materials. The success of the barley program and its scaling out to other crops created a need to address these questions, reinforced by growing international awareness and pressure to deal with them.

A similar process occurred in other cases, e.g. Nepal and China in particular. It was only after Nepal became a party to the CBD that the government and some NGOs started to discuss the importance of mainstreaming ABS issues in national policies. Similarly, following the country's engagement in the FAO's Commission on Plant Genetic Resources, national-level discussions were held to undertake initiatives for the conservation, management and use of plant genetic resources and, in this process, seek options to address farmers' concerns. NGOs, such as LI-BIRD and SAWTEE, made use of this policy space to bring local perspectives and interests to the table. Conversely, international concerns and issues were also introduced into local-level discussions and reflections.

As mentioned above, in most cases, farmers and indigenous communities have their own ideas, interests and practices concerning recognition and ABS, but they are often not expressed in formal ABS language. They are also usually maintained, transferred and adapted tacitly, that is, not written down, although some changes are occurring, in part due to the sort of development initiatives described in the case studies. They are often based on collective identities and forms of reciprocity, although in recent years these have come under strong pressure from privatization and commercialization forces. Farmers all over the world continue to rely heavily on informal seed systems, for example, in which a variety of modes of non-monetary and monetary exchange exist and through which recognition (e.g. farmers as expert seed producers), access (e.g. through biodiversity fairs) and benefit sharing (e.g. newly developed varieties are given away as gifts to neighbors to be tested) take shape. However, there are few societal incentives for farmers to maintain the local seed system other than for their own good, while, almost everywhere, breeders can obtain germplasm from farmers' fields for free. Farmers have no or little control over their genetic resources. Given that there is no societal compensation for farmers' conservation efforts, farmers' awareness about the wider importance of genetic diversity conservation has remained relatively weak. This is beginning to change now, though, as the cases demonstrate.

Recognition and valuation

How can local and indigenous knowledge and practices be acknowledged, recognized and valued? How can the principles of prior informed consent and mutually agreed terms (e.g. in the case of model agreements), including the settlement of possible disputes and remedies and arbitration, be respected?

The CBD, and Article 8(j) in particular, triggered a series of policy processes at national and international levels that seek to offer legal protection to traditional knowledge, including agro-ecological practices. These processes have also helped to revalue to some extent (from economic and legal perspectives more than from social or political ones) the role and importance of traditional knowledge, long acknowledged in social and agricultural disciplines and within local and indigenous contexts.

The intellectual input of communities into bioprospecting and breeding processes, in particular, and not only during the early stages of research and development, has proved critical in the production of new goods and services in a wide range of industries, including pharmaceuticals, biotechnology and agro-industry. PPB has also progressed as communities have a say in decision-making and defining their production priorities and needs. The value of their traditional knowledge has resulted in policy frameworks that now require prior informed consent (PIC) and the establishment of mutually agreed terms (MATs) before allowing access to and use of traditional knowledge.

PIC refers to consent by communities and farmers to the use of their knowledge and resources, based on a well-informed and timely process in which their decisions are based on appropriate data and information provided by potential users of the traditional knowledge and resources. PIC is perceived to be a means to redress the asymmetrical relationship between those seeking access (usually research institutions) and the holders or custodians of knowledge and resources. One of the main practical difficulties (and this was to some extent perceived by Jordan breeders and farmers) is identification of who precisely is entitled to grant PIC and agree to terms: a farmer, a group of farmers or communities? This becomes complicated when farmers and communities share resources and traditional knowledge, which is often the case. MATs, on the other hand, refer to a negotiation phase during which communities and farmers discuss and agree on how their seeds and materials may be accessed and used, by whom, under what conditions and for what specific purposes. MATs are a means of responding to the participatory principle that governs PPB.

Within this context, local knowledge and practice are diverse and constantly changing. Farmers and indigenous communities adapt to new conditions, often through research and development initiatives such as those highlighted in this book. Farmers all over the world continue to conserve and manage landraces in their local seed systems, which are under increasing pressure from market forces (e.g. in China, both government agencies and private-sector businesses are staging campaigns to sell hybrid seeds). For many researchers, the process of working with and learning from farmers has been one of awakening. As the Cuba case illustrates, although professional plant breeders faced a difficult economic situation after the country had to stand on its own feet, they continued their old, top-down approach believing that the best solution for all the problems in agriculture and plant breeding was "simple" technology substitution. A number of pioneers brought about change inspired by PPB experiences from elsewhere, based on the recognition that farmers are capable of experimentation and innovation and that, through joint efforts, perhaps better solutions could be found. PPB starts with recognition of farmers' knowledge and expertise, and includes the interest to build on it and strengthen it.

In Syria, scientists realized that users' participation in technology development may, in fact, increase its probability of success. In Jordan, PPB resulted in a dramatic change in attitude and behavior among breeders. They came to acknowledge and appreciate the knowledge and skills of farmers, and began to look for ways to build on their expertise. They also became aware that benefits are not just the final products of the breeding process (i.e. improved and released varieties), but that sharing of knowledge and experience is also a form of benefit sharing, leading to new insights, new experiences, new diversity and the step-wise improvement of farmers' crops and seeds. This was a major discovery and an important opening up of the conventional approach and system.

Maize research in China is well organized and has produced good results, but it has been carried out mainly in favorable production regions. Less favorable regions have not been served well. This has been partly because those involved in traditional plant-breeding science assume that farmers are less knowledgeable than breeders, that selection must be done under optimum conditions, that cultivars must be genetically uniform and widely adaptable over large geographic areas, and that landraces and open-pollinated varieties must be replaced by high-yielding varieties to ensure national food security. Such issues as biodiversity, farmers' diverse livelihoods and their contribution to crop improvement have been largely ignored.

PIC remains very much a novel idea and practice. None of the cases had a formal form of PIC at the beginning of the initiative, although they all implicitly accepted the principle, and some (Nepal and, more recently, China) formalized PIC later. In Nepal, communities were trained to document the genetic resources and associated traditional knowledge in a community biodiversity register (CBR). If the CBRs are recognized in policy and legal frameworks as certification by the custodians of this information and a national CBR is compiled, it will facilitate the process of bioprospecting, provide the basis for ownership of genetic resources and associated traditional knowledge, and specify the community to be involved in providing PIC. The recently signed novel ABS agreements in China are based on both PIC and MATs. They represent an inspiring example for other countries. In China, the team proposes that, within the existing legislation, farmers' rights can be protected through prior agreement and formal contracting between "parties" concerning the benefits to be shared. The contributions of farmers can be determined in various ways. The China case suggests that the contracting process could best be facilitated by an impartial third party. In addition to their role in providing germplasm, farmers' efforts in the PPB process should also be reflected in the contract.

Third parties may also be effective in dispute settlement. Right now, as most of the cases indicate, farmers have little or no recourse if a dispute arises. This issue merits further attention and research.

At the national level, the Andean Community was the first regional bloc to adopt a comprehensive policy and legal framework regarding access to genetic resources and the protection of traditional knowledge, as a pioneering step in implementing the equity and fairness principles of the CBD. Decision 391 of the Andean Community on a Common Regime on Access to Genetic Resources (1996) regulates who may have access to the region's genetic resources and under what conditions. It also sets general obligations for the recognition and protection of traditional knowledge. The whole Andean process to develop ABS and traditional knowledge frameworks arose from prior discussions regarding a regime to protect plant breeders' rights and concerns over access to and use of native and wild genetic materials.

This contrasts with Honduras where, just as farmers' seeds are unprotected under DR-CAFTA, so too is indigenous knowledge. Explicit clearance for the use of traditional knowledge or seed varieties by a patent applicant does not have to be provided; nor is the location of origin or an arrangement for benefit sharing between the applicant and the knowledge or seed holders required. In China, there are formal public registration systems for germplasm at both provincial and national levels; however, the current "passport" information for germplasm mainly focuses on genetic and geographic information, and lacks socioeconomic and cultural information about farmers and local communities. The custodians of genetic resources are treated as if they do not exist or do not matter in the formal system. To address this issue, the team proposes improving the current registration system to place more emphasis on farmers' and communities' rights, recognizing their crucial roles in maintaining agricultural biodiversity in the field through both individual and collective efforts.

Roles and responsibilities

How can the roles and responsibilities and the forms of participation of right-holders and stakeholders be defined (e.g. through formal or informal codes of conduct)?

Existing ABS policy and legal frameworks seek to organize how and under what conditions various actors involved in, for example, plant breeding, participate and engage in the breeding process or become involved in bioprospecting activities. This ranges from how breeding materials are obtained from in-situ or ex-situ sources to how benefits should be shared throughout the breeding cycle. One common element in almost all frameworks, including those in Peru, Nepal and, more broadly, China, is that some form of government permit is required to access and use materials.

Participatory approaches focus on meaningful, fair and iterative interaction. From PPB experiences around the world, we know this requires much effort. Those who take the initiative in practicing PPB need to pay special attention to:

- getting to know the various people involved, building trust and understanding, and respecting different (and sometimes, initially opposing) perspectives, interests and expertise
- acknowledging personal, social and institutional constraints to collaboration
- communicating clearly and in a timely manner
- finding common ground through discussion, reflection and negotiation
- defining tasks to be accomplished and agreeing on who will do what and when up front, e.g. setting objectives, selecting germplasm materials, choosing breeding/propagation/selection methods, selecting sites where the research will be carried out, identifying the type of end-product to be produced, and the means by which the product(s) will be distributed (i.e. benefit sharing).

For a more detailed discussion, see Vernooy *et al.* (2009).

Enhancement of the capacity of government as well as community institutions to develop the skills listed above remains a major task in all cases. In practical terms, it includes; legal access to breeding materials (if legislation is in place); registration of genetic resources and traditional knowledge; management of PIC; distribution, use and mobilization of benefit-sharing funds; and negotiation of terms with third parties. Some progress is being made, but there is still a long way to go.

The ability of national regulatory frameworks to support new roles and responsibilities varies greatly as the cases indicate. For example, although Jordan has adopted a comprehensive framework of agricultural policies and laws, ABS issues, especially in relation to PPB, have not yet been dealt with in a clear, concise and operational manner. The ABS team has made a start by identifying key issues in relation to the various elements of PPB, but the general lack of knowledge among researchers, policymakers and farmers has been a challenge. The Honduras case is an example of how difficult it is to change ingrained institutional (research) practices.

An important point arising from the cases is the role of the farmer and indigenous community organizations. The Nepal case makes the strongest argument for this role by saying that the establishment of a representative institution of farmers with a mandate for conservation and sustainable use of genetic resources and traditional knowledge is a *prerequisite* for protecting the rights of a community during implementation of any ABS regime. In Nepal, methods employed to strengthen farmers' organizational capacities included village-level workshops, folk song competitions, biodiversity fairs, rural street dramas and farmers' training in ABS and their right to contribute to this process of local organization. Honduras, Cuba and China are other examples where dynamic farmer organization processes are occurring.

Rights

What are the means to ensure respect for and conserve and strengthen indigenous/ local knowledge, customary practices and innovations? How should questions of intellectual property rights and ownership of genetic resources and related knowledge be dealt with? What are appropriate incentives and how can they be used?

ABS policies and laws and a growing number of frameworks designed to protect traditional knowledge are paving the way to new thinking about how to preserve ancestral traditional knowledge and ensure compensation when it is used. Once again, finding appropriate synergies and connections between customary practices and formal, state law is proving difficult. On one hand, these frameworks are focusing on traditional knowledge as a "commodity" and its potential practical use. This obscures the fact that traditional knowledge is an element in a cultural context where direct or potential use is but one of the many ways in which indigenous peoples and communities (including farmers) access and give meaning to genetic resources. For many local communities around the world, genetic resources also have religious, ethical, spiritual and sociocultural meaning and value. Use and possible appropriation of genetic resources are alien to these values. At the same time, communities are advocating further respect and recognition of local practices, where, for example, biocultural protocols are designed, a priori, to set the standard under which traditional knowledge may be accessed or used.

As mentioned, ownership and, more so, intellectual property rights are often detached from a sociocultural context where sharing and freely exchanging resources (seeds) and traditional knowledge is the norm. In the case of PPB in China and Jordan, in particular, ownership and rights questions have arisen. The answers to these questions will define the effective beneficiaries and the future incentives for PPB. Enclosing a traditional common good (a seed or variety) will affect perceptions and possibly curtail future innovation at the local level and harm partnerships with research institutions.

Sharing knowledge and expertise is a concrete and important way to share benefits, and all the cases in this book emphasize this. As the Syria case argues, combining farmers' knowledge with that of professional breeders enables farmers to benefit from their contributions to the global genetic pool, for example, in added value to their crops, improved livelihoods and increased income. In Honduras, throughout the participatory breeding process, farmers received extensive agronomic support from FIPAH, the NGO facilitating the CIAL process. None of the farmers had segregated material before and had to learn how to select for characteristics that might vary from one generation to the next. This was also a new process for FIPAH and a good deal of mutual learning took place.

The Potato Park, cited in the Peru case, was created to protect the biocultural collective heritage of *campesino* communities. This implies a series of activities, including protecting traditional knowledge, preventing biopiracy, developing local biodiversity and traditional knowledge registers, promoting ecotourism, repatriating lost crops (with support from the International Potato Center) and promoting the sale of local products including soaps, shampoos and medicinal plants. The concept of the Potato Park is to provide communities with a development option based on their own needs and interests and using market forces to satisfy these interests. Critically important are the cultural, spiritual and ancestral customs as elements guiding livelihoods and activities in the park. The park is an example of an "agro-biodiversity zone," which has gained formal recognition in Peru.

The concept of farmers' rights, as promoted by the ITPGRFA, has made inroads in some countries (most notably in Nepal and, to some extent, Peru and Jordan), but it remains a challenge to integrate and operationalize the three basic rights referred to in the treaty in national policy and laws: protection of traditional knowledge relevant to PGRFA; equitable sharing of benefits arising from the use of PGRFA; and participation in decisions, at the national level, on matters related to the conservation and sustainable use of PGRFA. In some countries, even acceptance of the concept is not that easy, given the political context (e.g. Honduras, Syria, Cuba and China). However, the case studies have explored a variety of ways to put the concept into practice. In Table 12.2, we summarize how

Table 12.2 Overview of farmers' rights in practice

Right	Practice
Protection of traditional knowledge relevant to PGRFA	*De facto* and formal legal recognition of collective forms of traditional knowledge and practices
	Use of prior informed consent for research and development initiatives
	Legal recognition of community biodiversity registers
	De facto acceptance and formal legal recognition of farmers as competent plant breeders and conservationists of biodiversity
	De facto and legal acceptance of farmers as bona fide seed producers
	De facto and legal acceptance of local forms of farmer and community organization (including agro-biodiversity zones)
Participation in benefit sharing	Capacity building among farmers and community members in a variety of areas related to crop improvement, conservation of biodiversity and rural livelihood improvement
	Involving women through special activities in crop improvement and other rural livelihood improvement efforts
	Improved (local) varieties; access to new breeding materials and related knowledge in hands of the formal sector
	Increased crop diversity; access to new resources for conserving biodiversity in situ
	Improved local seed systems; establishment of local seed banks; access to new channels for the outflow and inflow of seeds
	New seed production and commercialization opportunities
	Royalties arising from crop improvement
	Access to local biodiversity fund
Participation in national decision making	Participation in local, regional, national and international policy workshops, seminars, conferences
	Indirect involvement through research and development organizations

the case studies give meaning to the concept of farmers' rights, explicitly or implicitly. As can be seen, it has been easier to put the first two rights into practice than the third. Farmers still do not have any direct role, as legitimate stakeholders, in national decision-making processes, whether related to general policy development or to specific measures, such as variety release policies.

In Honduras, Nepal and Peru, the risk of biopiracy seems to be a serious concern for small farmers. Specific traits in the landraces that farmers have conserved or improved through PPB might become materials protected under UPOV-91 or one or more national patent regimes. In Honduras, the case study authors fear that should a law for the protection of plant varieties eventually pass, small farmers are unlikely to enjoy much protection from farmers' rights regulations; this situation may not be an exception.

Science is evolving rapidly in some fields, and policies cannot always catch up. Over the past decade or so, advances have been changing how research and development related to biodiversity components takes place. Bioinformatics, genomics, proteomics, synthetic biology and genetic engineering have revolutionized how encoded natural information in genes and other molecular structures can be read, manipulated and transformed into useful products in almost all sectors of human activities. As a result, decades-old legal frameworks and even current templates cannot be appropriately applied to these new research paradigms and have, therefore, to an important extent, become outdated and, more troubling, inapplicable. Rights over genetic resources acquire new meaning in light of these developments. More research on their impact seems warranted.

Mechanisms and incentives for ABS

How can feasible ABS mechanisms, both formal and informal, be designed, implemented and monitored? How can conflicts between local-level ABS priorities and national/international interests be avoided? How can existing conflicts be resolved? How can conditions be created to reduce future conflicts?

Given the limited implementation of ABS policy and legal frameworks worldwide, there has been little opportunity to test monitoring mechanisms intended to ensure that benefit sharing takes place. Some ideas have been proposed, from tracking flows of resources to demanding strong reporting requirements along the research and development chain (Ruiz and Lapeña 2007).

The teams associated with the case studies in this book have experimented with a wide array of mechanisms covering the various steps involved in genetic conservation and crop improvement. The Syria case argues, echoing other PPB cases, that PPB is an effective way to generate and share benefits. Yield increases brought about by PPB are a concrete way to improve livelihoods. The Syria case also shows that no matter how many varieties are released from the formal system and no matter how much greater their yields are, farmers in marginal environments will not adopt them unless they have participated in the selection process. This makes PPB a particularly important tool for benefit sharing. Analysis of the farm-level benefits and costs of barley production showed that the participation of farmers in the breeding program does not mean higher production costs. Farmers who adopt varieties bred through PPB often pay higher input costs, but gain higher net returns. As the case study argues, in addition to economic benefits, participating farmers gain in terms of increased knowledge of barley production and variety selection and from their collaboration with scientists and other farmers. This type of non-monetary benefit is critical and demonstrates the overall importance of PPB and farmer participation.

However, all issues are not as well defined as they could be. Jordan may benefit from a national law on farmers' rights, but it has not yet been feasible to define clear ABS principles among PPB farmers, in particular concerning seed multiplication and distribution. Farmers *do* have an interpretation of benefit sharing. Some of them produced seeds and distributed some free to other farmer participants in the PPB research. One farmer sold his new variety and recorded the names of farmers who bought seeds to be able to track the diffusion process. However, how to translate this reality into adequate policy and legislation remains a challenge. Currently, Nepalese draft legislation accepts farmers as breeders of new varieties, but there is still no mechanism for providing incentives to farmer breeders. The same is true in Honduras and China.

How real or hypothetical benefits from PPB should be shared is not easy to determine, however. The Honduras case is the most outspoken about this. The authors conclude that it is clear from farmers' allocation of the hypothetical benefits from PPB that they are not prepared to accord the breeder a significant portion of the benefits, even when they are using breeder materials: the breeder, who is generally out of sight, is largely out of mind. Instead, farmers regard their labor and skills as the main ingredients of PPB; human resources appear to be more important to them than rights over local germplasm. Thus, the longer and more complex the process of PPB, the more farmers are likely to feel they have rights over the benefits ensuing from it, independent of where the germplasm originated. From the breeders' perspective, however, farmer selection/breeding involving breeder materials is more likely to be viewed as validation of their skills than as farmer creativity. This is particularly true if breeders have little opportunity to witness the skills and effort that farmers put into the breeding process. Just as farmers remain unaware of the resources (both human and financial) invested in plant breeding, breeders who rarely stray from the experimental station are likely similarly uninformed. This contrasts with the China case, where an agreement has been reached based on acceptance of joint efforts in terms of process and outcomes to share benefits through a collective mechanism (community fund) in support of community efforts.

Biodiversity or seed fairs (Cuba, Nepal and China) are important venues for the exchange of knowledge, experience and seeds. In Cuba, they were used to start a process of seed diffusion and large-scale, on-farm testing of new lines and varieties. They are a much appreciated way for the formal sector to "open doors" to benefit farmers, who have embraced the fairs with open arms. What is more, farmers are replicating fairs at the local level, organizing and financing them largely on their own. In China and Nepal, fairs, above, all function as exchange platforms.

Seed production has good potential for generating monetary benefits. But producing seed according to national rules and regulations is not always easy for small farmers. In Cuba, this went smoothly. In the case of the Felo variety of maize, seed multiplication and continued selection proceeded very well, and the variety has been officially registered. None of the stakeholders has raised obstacles to limit access to the new variety or block the generation and sharing of benefits (the cooperative of which farmer Felo is a member has become a seed producer). The process leading to Felo created new understanding, attitudes and behavior concerning ABS, which were carried forward by farmers and breeders alike. Elsewhere, seed production has encountered technical, managerial and regulatory obstacles, as the cases of Jordan, China and Honduras illustrate.

Value can also be added by applying a "geographic indication" designation to certain crops to distinguish specific local high-quality seeds and derived products, such as the wax maize grown in some parts of Guangxi. All the farmers within the area may join and benefit from collective production, and the certification could be applied and managed by a local farmer organization or a producer group with support from its administrative village. The Nepalese team is also exploring such an arrangement.

A community-based biodiversity management (CBM) fund is a way to share benefits acquired from common resources in villages. In Nepal, every household in a village development committee is eligible for a loan from the fund. Experience has shown that a CBM fund can contribute greatly to sustainable biodiversity management in a community. After seeing how benefits can be distributed successfully through CBM funds, the project team is now investigating whether money acquired from the use of community genetic resources can go directly into the fund and be used equitably for biodiversity management and community welfare. Supporting legislation is needed, and the team is lobbying for this. The China team is experimenting with a similar mechanism.

The China team argues that not only is support needed at the local level, but a national research program on landrace conservation and improvement should also be set up as part of the working agenda of all plant breeding institutes in the country. Efforts of breeders in this area should also be recognized and evaluated in institutes' annual performance reviews. Another option is to set up a national registration system for open-pollinated varieties (OPVs), including landraces, traditional varieties and farmer-improved OPVs, in parallel with the "new varieties" protected by law. Within this system, the diversity of plant genetic resources can be captured and the contribution of breeders (both farmer breeders and formal-sector breeders) can be recognized.

Who benefits? Gender and other social variables do matter. In most cases, participatory fieldwork has been used to understand gender-based differences in agronomic management, crop preferences and needs (see the Syria case for a discussion). As a result, PPB activities are now organized in ways that facilitate the involvement of women farmers. This is done by coordinating the events directly with women as well as collaborating with local institutions and by creating women-only spaces. The team tries to respect local sensitivities, particularly with regard to the participation of young female farmers in public events. In China, most farmers are women, and this has led the team to pay careful attention to gender issues from the start. However, not all project teams have been so aware of gender differences, suggesting that more work is needed.

Effective implementation will be the ultimate test of any ABS regime. Peru, which developed a *sui generis* policy early on, and competent authorities (the National Institution for the Defense of Competition and Intellectual Property and the Ministry of the Environment) have made strong efforts to implement regulations. However, there is still much to do in terms of strengthening institutional capacities to apply norms and monitor their implementation—in Peru and elsewhere.

The way forward

Constructing ABS and traditional knowledge protection frameworks is a relatively new process, compared with the development and consolidation of the intellectual property rights system which has taken more than a century to evolve. Much as countries (including Peru, Nepal, China) have rapidly developed their particular ABS and traditional knowledge laws and regulations, it is clear from practice that much more interaction with farmers (and communities) is required to ensure that these policies and norms reflect reality and do not rapidly become obsolete. Benefit sharing, PIC and MAT principles, although they are becoming recognized as formal concepts, are often difficult to understand and apply at local levels where more traditional and customary practices take place, based on collective identities, knowledge and practices. This points to the need to continue searching for alternative ABS policy and legal options, beyond the still narrow approach that prevails at the moment and is reflected in the Nagoya Protocol.

At a minimum, local benefit sharing options and practices, especially in agriculture, should be respected and used as a basis for the development of national or international policy and legal frameworks instead of "squeezing" them into narrowly defined options. Understanding local context is also key to ensuring harmonious implementation. ABS discussions can cause confusion, deter openness and affect confidence-building processes among farmers and other actors.

Building dynamic and supportive institutional and formal legal structures from the bottom up, especially in terms of access to resources and traditional knowledge and the resulting benefit sharing, is necessary to ensure practical results. Reflecting on local initiatives and developing biocultural protocols or even codes of conduct with direct input from farmers and communities could create synergies among local, national and international levels and help implementation processes. More local policy experimentation (as highlighted in the case studies) seems warranted to provide national and international decision-makers with more grounded examples of how ABS can work. Improving communication between local-level practitioners and national and international decision-makers is another major necessity for improving ABS policymaking and law-making.

The cases provide concrete examples of the key policy measures contained in the ITPGRFA: promotion of diverse farming systems, including the use of local crops, varieties and underutilized species; support for research that enhances biological diversity; broadening of the genetic base of crops in situ and ex situ; creation of stronger links to plant breeding and agricultural development; promotion of PPB; and review/adjustment of breeding strategies and regulations concerning variety release and seed distribution. They have done this mostly with international donor support. National governments could set up funds to permit wider adoption and adaptation of these measures allowing more farmer communities to experiment and reap the benefits. The organizations that have acquired some expertise in this field (such as the protagonists in this book) could be invited to play an advisory role in setting up appropriate mechanisms and support the implementation of new initiatives. They could also strengthen or build national networking activities.

New technologies and new research and development approaches, such as bioinformatics and genomics, are dramatically changing the ways in which research and product development take place. How actors engage in these processes, how biological materials are accessed and provided, how innovation is being protected, the role of intellectual property rights and the informational nature and widespread distribution of genes and biodiversity components, among other variables, are not being addressed nor appropriately discussed in policy and legal forums. As a result, existing policies and laws (including the Nagoya Protocol) seem very much detached from these trends. It is high time for them to catch up (see the Epilogue for a succinct elaboration of this point of view).

PPB efforts offer a unique example of how benefit sharing takes place in practice, with or without an overarching ABS or traditional knowledge policy or legal framework in place. Examples from Syria, Jordan, Honduras, China, Nepal and Cuba demonstrate that actors in the research and development chain participate in and generate benefits that are distributed according to a wide range of criteria that often go beyond narrowly defined policy and legal guidelines. These examples ought to be better heard and taken into consideration at national and international levels.

References

CBD (Convention of Biological Diversity) (n.d.) Country profiles. CBD website. Available at: www.cbd.int/countries (accessed 30 March 2011).

COP 10 (Tenth Conference of the Parties) (2010) Access to genetic resources and the fair and equitable sharing of benefits arising from their utilization (advanced unedited version). Convention on Biological Diversity website. Available at: www.cbd.int/cop/cop-10/doc/advance-final-unedited-texts/advance-unedited-version-ABS-Protocol-footnote-en.doc (accessed 30 March 2011).

FAO (Food and Agriculture Organization of the United Nations) (2009) International treaty on plant genetic resources for food and agriculture. FAO, Rome, Italy. Available at: ftp://ftp.fao.org/docrep/fao/011/i0510e/i0510e.pdf (accessed 30 March 2011).

FAO (Food and Agriculture Organization) (n.d.) International treaty on plant genetic resources for food and agriculture. FAO, Rome, Italy. Available at: www.fao.org/Legal/treaties/033s-e.htm (accessed 30 March 2011).

Ruiz, M. and Lapeña, I. (2007) A moving target: genetic resources and options for tracking and monitoring their international flows. IUCN Environmental Policy and Law paper 67/3. IUCN, Gland, Switzerland.

Vernooy, R., Shrestha, P., Ceccarelli, S., Ríos, H., Song, Y. and Humphries, S. (2009) Towards new roles, responsibilities and rules: the case of participatory plant breeding. In S. Ceccarelli, E. Guimaraes and E. Weltzien-Rattunde (eds.), *Plant breeding and farmer participation.* Food and Agriculture Organization of the United Nations, Rome, Italy. pp. 613–28.

Epilogue

Architecture by committee and the conceptual integrity of the Nagoya Protocol*

Joseph Henry Vogel

"Architecture by committee" produces appalling esthetics. Anyone familiar with university campuses can conjure up the image of a hideous building amid esthetically pleasing ones. An analogy exists for the United Nations Convention on Biological Diversity (CBD) and the building of the international regime on access and benefit sharing (ABS) (COP 10 2010). I will call it "policymaking by consensus." Delegations in nine working groups labored for years to draft a protocol for the Tenth Conference of the Parties (COP 10) which was held in Nagoya, Japan, 18–29 October 2010. Unfortunately, the experts in the delegations did not constitute an independent authority immune to political pressure; allegiance in COP 10 and previous working groups has been to the delegation and not to the expertise. Whatever conceptual integrity may have existed was expunged as the bracketed text began to lose the brackets. Although policymaking by consensus seems democratic, it is anything but. Coherence is effectively denied everyone.

The economics of information can provide an unencumbered vision for ABS and the case studies of this volume. Inspired by Theodosius Dobzhansky's (1973) seminal article "Nothing in biology makes sense except in the light of evolution," I would say that "Nothing in the international regime makes sense except in the light of the economics of information." Moreover, "[s]een in the light of [the economics of information, ABS] is, perhaps, intellectually the most satisfying and inspiring [objective of the CBD]. Without that light [ABS] becomes a pile of sundry facts some of them interesting or curious but making no meaningful picture as a whole" (Dobzhansky 1973: 129).

I have selected two articles from the Nagoya Protocol that may show the reader how to begin. Article 2(e) establishes a key definition that has been under discussion since ABS working group 5 (Appleton *et al.* 2007). Four working groups and two COPs later, the delegations have agreed that:

> "Derivative" means a naturally occurring biochemical compound resulting from the genetic expression or metabolism of biological or genetic resources, even if it does not contain functional units of heredity.
>
> (COP 10 2010)

I will again plumb the wisdom of evolutionists. The illustrious Richard Dawkins (2008) describes genes as "pure information" and I would only qualify that characterization with the adjective "natural." The question arises: could natural information be used in patentable research and development (R&D) but fall outside the definition of "derivative" established in Article 2(e)?

"Of the division of labour" was the first chapter of Adam Smith's (1776) *An inquiry into the nature and causes of the wealth of nations.* As a professor of economics, I assigned to my graduate students the task of finding examples of natural information that resulted in patents but did not involve any biochemical compound. The assignment proved ridiculously easy. Students found not only a hugely successful patent, viz., Velcro® (US patent 2,717,437 filed on 15 October 1952 (Velcro n.d.)), but also a whole field of R&D outside the black letter of Article 2(e). "Biomimicry ... [is] a new discipline that studies nature's best ideas and then imitates these designs and processes to solve human problems" (Biomimicry Institute n.d.a). One infers from the website of the Biomimicry Institute (n.d.b) that intellectual property is the enabler. Under the Nagoya Protocol, no benefit from any patented biomimicry would have to be shared with any country of origin.

Were such anomalies identified in the run-up to Nagoya? The question needs no division of labor. In 2007, I published a comprehensive article in the International Union for the Conservation of Nature ABS series which is freely available online in English, French and Spanish (Vogel 2007a). In Table 1, "Tilted playing fields in the hyperspace of ABS," the columns identify 16 distortions in light of the general theory of second best and the economics of information. Distortion number 13 is "symbolic phenotypic expressions ... for patents on designs inspired from nature"—in other words, biomimicry. But even in 2007, such anomalies were not new to the CBD literati. *In Genes for sale* (Vogel 1994: 43), the textbox "Monkey know-how" describes how "[t]he remarkable Jane Goodall ... documented that ill chimpanzees pick leaves of plants known to possess therapeutic effects." For patents arising from biomimicry or non-human culture, no ABS agreement would be necessary under Article 2(e), as no "biochemical compound" would be obtained in R&D.

My students may accuse the working groups and delegates to the COPs of not doing their homework, but homework is the wrong metaphor. Lack of due diligence is more appropriate, given that the whole CBD process has been "overly legalistic," as cogently argued by the editors of this volume. Anomalies, such as biomimicry and non-human culture, will eventually gain traction in the public consciousness.

What will be the response of the newly formed Intergovernmental Committee to the Nagoya Protocol? A hopeful insight can be gleaned from Thomas Kuhn (1970: 78):

> They will devise numerous articulations and *ad hoc* modifications of their theory in order to eliminate any apparent conflict. If, therefore, these epistemological counterinstances are to constitute more than a minor irritant, that will be because they help to permit the emergence of a new and different analysis of science within which they are no longer a source of trouble.

On careful rereading of the above quote, I doubt that the anomalies alone will give way to a paradigm shift. Policymaking by consensus is not a "theory" that can "permit the emergence of a new and different analysis of science." It is a *modus operandi*. We are not confronted with Kuhnian paradigm choice within science. We are confronted with a more fundamental choice between non-scientific and scientific approaches to resource allocation.[1] Among the latter, the economics of information is now in the Kuhnian stage of normal science puzzle-solving. Nobel Memorial Prizes have been awarded and ABS is merely low-hanging fruit for the plodding economist (Sveriges Riksbank Prize 2001). The fruit metaphor returns me to the primatology literature and utter discouragement. Comparing human politics with that of the bonobos, Frans de Waal (2005) notes that "the persuasive power of logic is surprisingly limited."

Other critics of the Nagoya Protocol have also lit on Article 2(e) and deployed logic with similar failure to persuade. Pat Roy Mooney, Executive Director of Erosion, Technology, and Concentration (ETC) spoke eloquently at various side events at COP 10. ETC's advice to the delegations:

> Parties should define *Derivative* as a digital sequence, biochemical compound, engineered organism or metabolic pathway resulting from the collection, genetic expression or metabolism of biological or genetic resources, even if they do not contain functional units of heredity.
>
> (ETC Group 2010; emphasis in original)

The reader should note well that every item in the ETC recommendation can be classified as natural information as well as the items forgotten, i.e. biomimicry and non-human culture. But even if ETC had included biomimicry and non-human culture, the laundry-list definition would still have been inferior to natural information. The reason why is subtle. Natural information is a category sufficiently capacious to allow inclusion of phenomena still unidentified. Kuhn (1970: 111) writes:

> Led by a new paradigm, scientists adopt new instruments and look in new places. Even more important, during revolutions scientists see new and different things when looking with familiar instruments in places they have looked before.

After a bit of daydreaming, I can imagine the existence of useful fractal designs in the crystals of thermal pools threatened by nearby geothermal exploration. A Google search quickly generates hits. Defining the object of ABS as natural information would include such crystals; the *ad hoc* definitions from ETC or Article 2(e) would not.

With the object of ABS redefined as natural information, I proceed to analyze Article 10 whose theme undergirds the case studies presented in this volume:

Global Multilateral Benefit-Sharing Mechanism
Parties shall consider the need for and modalities of a global multilateral benefit-sharing mechanism to address the fair and equitable sharing of benefits derived from the utilisation of genetic resources and traditional

knowledge associated with genetic resources that occur in transboundary situations or for which it is not possible to grant or obtain prior informed consent. The benefits shared by users of genetic resources and traditional knowledge associated with genetic resources through this mechanism shall be used to support the conservation of biological diversity and the sustainable use of its components globally.

(COP 10 2010)

The opening words "Parties shall consider ..." means that ratified parties need only *contemplate* a global multilateral benefit-sharing mechanism; they need not subscribe to one. In popular parlance, such phrases are "weasel words" and the protocol is replete with them (Wasserman and Hausrath 2006). Nevertheless, the first sentence of Article 10 does recognize that the object of ABS is diffused across taxa and political boundaries.

Pity it ends with the blatantly false premise that "it is not possible to grant or obtain prior informed consent." Not possible? Inasmuch as ratified parties and traditional communities are sovereign over their genetic resources and undisclosed traditional knowledge, diffusion does not matter one whit in terms of granting or obtaining prior informed consent. However, diffusion does matter a great deal for fair and equitable benefit sharing. Biotech companies will comparison-shop among countries and communities and drive the price of access down to the marginal cost of collection, which is essentially nothing. The outcome can only be averted if the Global Multilateral Benefit-Sharing Mechanism sets the price of access and denies ratified parties the power to negotiate bilaterally.[2] I have advocated such a mechanism for a long time and elaborated the institutional details in *The biodiversity cartel: transforming traditional knowledge into trade secrets* (Vogel 2000). Again, no due diligence is evident in the drafting of Article 10.

The other sentence in Article 10 is also frustrating for anyone who insists on conceptual integrity. Cartelization demands that the proceeds go to the treasuries of the countries of origin. But policymaking by consensus has resulted in the proceeds going to "the conservation of biological diversity and the sustainable use of its components globally." To the extent that such activities are fungible, earmarking will simply displace funding that would have been allocated anyway.[3] The most outlandish example of displacement would be a future COP to the Nagoya Protocol. By Article 10, the mechanism can become the source of COP funding! I do not think that is what Adam Smith had in mind when he celebrated the virtues of pursuing one's own interest.

A biodiversity cartel would distribute the benefits proportional to the geographic size of the habitats in the transboundary countries. This seems to be the simplest and fairest rule. For species so widely distributed that the costs of determining the habitat outstrip the royalties collected, the sum "should be used to diminish the fixed costs of the Gargantuan Database" (Vogel 1992). I published those words in 1992 long before anyone imagined that the International Barcode of Life (iBOL) would emerge as the gargantuan database and enable the cartel (see Vernooy *et al.* 2010).

The unencumbered vision of ABS that I have outlined in this Epilogue seems to extricate COP delegates from any substantive policymaking. Letting such an impression stand would also be lacking due diligence—this time, on my part. In a *Foreign Affairs* article entitled "Is government too political?" the economist, Alan S. Blinder (1997) explains how *both* efficiency and equity can be realized by circumscribing the scope of politics. Blinder implores the public to contemplate "what things the government should and should not be doing ... different arrangements for governance draw the line between political and technocratic decisions in different places and every society must choose where the line should fall." He offers numerous examples, but the CBD is not among them; its absence confirms Kuhn's (1970) insight about new paradigms proving their worth through applications to unintended areas.

Speaking in the most general terms, Blinder would delegate the technical aspects of any policy to the experts and leave the value judgments for the politicians. Let me apply that wisdom to two points about the biodiversity cartel: the issue of the royalty rate and the relative status of "confidential business information" versus "transparency." They turn out to be interrelated and both pivot on value judgments. To public queries about the royalty rate in consummated material transfer agreements (MTAs), industry routinely invokes the mantra of "confidential business information." No wonder: when the royalty rates are leaked, we discover percentages of about half of one percent (I dare not write 0.5% as a reader may misread the number as 5%; Edmonds Institute 1999). However, when one questions the public about what they think would be a fair and equitable royalty rate, a typical response is 50% (NEF 2005). So, we are talking about a breach of two orders of magnitude. Technocrats should bow out of the negotiation between 0.5% and 50% and let the delegates do battle in the COPs.

The dilemma among rights in conflict also defies technical expertise. It is a value judgment whose nature is binary, unlike the royalty rate which is a continuum. Does the "confidential business information" implicit in bilateral "MTAs" trump the "transparency" of MTAs or does "transparency" trump "confidential business information'? Tellingly, MTAs appears 25 times in the Nagoya Protocol and transparency appears just twice. The behaviorist B.F. Skinner would be proud; the operant conditioning techniques he pioneered for rats are also effective in humans. The issue of transparency in MTAs is not yet on the table. Inasmuch as no one can resolve a priori the supremacy of rights, I have suggested a venue to vet them in *The museum of bioprospecting, intellectual property, and the public domain: a place, a process, a philosophy* (Vogel 2010).

I began this Epilogue with an abstract analogy about architecture by committee and policymaking by consensus. I will close with a concrete example, pardon the pun, from my own work environment, the University of Puerto Rico. The flagship campus at Río Piedras boasts a lovely quadrangle designed in 1936 by the architect William Parsons. The original plan was in the style of the Spanish Renaissance (Alumni n.d.). Lush shade trees provide welcome refuge from the tropical sun. Decades later, the conceptual integrity of the quadrangle was violated by the construction of the Faculty Senate. The building looks like a morgue. It sits at the base of the beautiful clock tower which is the focal point of the quadrangle and emblem of the university. Plans to demolish the Faculty Senate have never been realized but I remain hopeful.

Notes

*   Indirect support for this work was provided by the United States National Science Foundation (IGERT grant 0801577) and the Australian Research Council (grant LX0881935). I would like to thank the following students who provided background research on anomalies to Article 2(e) of the Nagoya Protocol: Nora Alvarez, Arelis Arocho, Norberto Quiñones-Vilche, Jeiger L. Medina Muñiz and Julio Miguel Santiago Ríos. Special thanks are extended to Barbara A. Hocking, Maritza Stanchich and Paul Baymon who carefully proofread the manuscript.

1   In "Reflecting financial and other incentives" (Vogel 2007a), I identify two schools of thought on the economics of bioprospecting. Track I is associated with cost–benefit analysis and attempts to compute the value of genetic resources for bioprospecting using probabilistic models of pharmaceutical discovery. Track II is associated with the *realpolitik* of creating a countervailing force to battle with vested interests which would dismantle limits on land use. Track III, not mentioned in the article, would be the open access movement. See Oldham (2009).

2   For standardizing the royalty rate across all species regardless of diffusion, see Vogel (2007b).

3   Regarding fungibility, see the example of the Bill and Melinda Gates Foundation in Vogel (2009).

References

Alumni (n.d.) Rehabilitación del Cuadrángulo Histórico y Torre de la UPRPR. Alumni, Universidad de Puerto Rico, Recinto de Río Piedras, San Juan, Puerto Rico. Available at: http://alumni.uprrp.edu/torre.asp.html (accessed 2 April 2011).

Appleton, A., Jinnah, S., Jonas, H., Jungcurt, S. and Schabus, N. (2007) ABS 5 highlights. *Earth Negotiations Bulletin*, 7(390). Available at: www.mail-archive.com/enb@lists.iisd.ca/msg00435.html (accessed 2 April 2011).

Biomimicry Institute (n.d.a) What is biomimicry? Biomimicry Institute, Missoula, Mont., USA. Available at: www.biomimicryinstitute.org/about-us/what-is-biomimicry.html (accessed 30 January 2011).

Biomimicry Institute (n.d.b) Biomimicry: a tool for innovation. Biomimicry Institute, Missoula, Mont., USA. Available at: www.biomimicryinstitute.org/about-us/biomimicry-a-tool-for-innovation.html (accessed 30 January 2011).

Blinder, A.S. (1997) Is government too political? *Foreign Affairs* 76(6): 116.

COP 10 (Tenth Conference of the Parties) (2010) Access to genetic resources and the fair and equitable sharing of benefits arising from their utilization. Secretariat of the Convention on Biological Diversity, Montreal, Canada. Available at: www.cbd.int/decision/cop/?id=12267 (accessed 2 April 2011).

Dawkins, Richard (2008) Life: a gene-centric view. Craig Venter and Richard Dawkins: a conversation in Munich. Website of Edge: The Third Culture. Available at: www.edge.org/documents/dawkins_venter_index.html (accessed 30 January 2011).

de Waal, F. (2005) *Our inner ape.* Riverhead Books, New York, NY, USA. p. 197.

Dobzhansky, T. (1973) Nothing in biology makes sense except in the light of evolution. *The American Biology Teacher* 35:125–9.

Edmonds Institute (1999) Mexico's genetic heritage sold for twenty times less than the US got in Yellowstone (press release, 28 September 1999). The Edmonds Institute, Edmonds, Washington, USA. Available at: www.biotech-info.net/genetic_heritage.html (accessed 2 April 2011).

ETC (Erosion, Technology, and Concentration) Group (2010) ETC Group briefing for CBD COP 10 in Nagoya: synthetic biology. Website of Hands Off Mother Earth. Available at: http://handsoffmotherearth.org/2010/10/etc-group-briefing-for-cbd-cop-10-in-nagoya-synthetic-biology/ (accessed 2 April 2011).

IISD (International Institute for Sustainable Development) (2007) ABS highlights: Tuesday 9 October 2007. *Earth Negotiations Bulletin* 9(390). Available at: www.iisd.ca/vol09/enb09390e.html (accessed 30 January 2011).

Kim, J. and Peacor, D.R. (2002) Crystal-size distributions of clays during episodic diagenesis: the Salton Sea geothermal system. *Clays and Clay Minerals* 50(3): 371–80. Available at: http://ccm.geoscienceworld.org/cgi/content/abstract/50/3/371 (accessed 2 April 2011).

Kuhn, T. (1970) *The structure of scientific revolutions* (2nd edn.). University of Chicago Press, Chicago, Ill., USA.

NEF (New Economics Foundation) (2005) *Behavioral economics: principles for policy makers* (principles 3 and 6). NEF, London, UK. Available at: www.neweconomics.org/sites/neweconomics.org/files/Behavioural_Economics_1.pdf (accessed 2 April 2011).

Oldham, P. (2009) An access and benefit-sharing commons? The role of commons/open source licenses in the international regime on access to genetic resources and benefit sharing. Initiative for the Prevention of Biopiracy, Research Documents, Year IV No. 11. Available at: http://ssrn.com/abstract=1438027 (accessed 2 April 2011).

Smith, Adam (1776) *An inquiry into the nature and causes of the wealth of nations.* W. Strachan and T. Cadell, London, UK.

Sveriges Riksbank Prize in Economic Sciences in Memory of Alfred Nobel (2001) George A. Akerlof, A. Michael Spence, Joseph E. Stiglitz. Website of the Nobel Prize. Available at: http://nobelprize.org/nobel_prizes/economics/laureates/2001/ (accessed 2 April 2011).

Velcro (n.d.) Innovation at work. Velcro, Manchester, NH, USA. Available at: www.velcro.com/index.php?page=innovation (accessed 30 January 2011).

Vernooy, R., Haribabu, E., Muller, M.R., Vogel, J.H., Hebert, P.D.N., Schindel, D.E., Shimura, J. and Singer, G.A.C. (2010) Barcoding life to conserve biological diversity: beyond the taxonomic imperative. *PLoS Biol* 8(7): e1000417 (13 July). Available at: www.plosbiology.org/article/info:doi/10.1371/journal.pbio.1000417 (accessed 2 April 2011).

Vogel, Joseph H. (1992) *Privatisation as a conservation policy.* CIRCUIT, Melbourne, Australia. p. 96.

—— (1994) *Genes for sale.* Oxford University Press, New York, NY, USA.

—— (ed.) (2000) *The biodiversity cartel: transforming traditional knowledge into trade secrets.* CARE, Quito, Ecuador.

—— (2007a) Reflecting financial and other incentives of the TMOIFGR: the biodiversity cartel. In M. Ruiz and I. Lapeña (eds.), *A moving target: genetic resources and options for tracking and monitoring their international flows.* International Union for the Conservation of Nature, Gland, Switzerland. pp. 47–74. Available at: http://data.iucn.org/dbtw-wpd/edocs/EPLP-067-3.pdf (accessed 30 January 2011).

—— (2007b) From the "tragedy of the commons" to the "tragedy of the commonplace": analysis and synthesis through the lens of economic theory. In C.R. McManis (ed.), *Biodiversity and the law: intellectual property, biotechnology and traditional knowledge.* Earthscan, London, UK. pp. 115–36.

—— (2009) *The economics of the Yasuní Initiative: climate change as if thermodynamics mattered.* Anthem Press, London, UK. p. 61.

—— (ed.) (2010) *The museum of bioprospecting, intellectual property, and the public domain: a place, a process, a philosophy.* Anthem Press, London, UK.

Wasserman, P. and Hausrath, D. (2006) *Weasel words: the American dictionary of doublespeak.* Capital Books, Herndon, Va., USA.

Index